The Luxor Obelisk

AND ITS VOYAGE TO PARIS

The Luxor Obelisk

AND ITS VOYAGE TO PARIS

APOLLINAIRE LEBAS

Introduced and translated by Bob Brier and Colette Fossez Sumner

The American University in Cairo Press
Cairo New York

Title page:
Caricature of Apollinaire Lebas
by Jean-Pierre Danton, 1839;
the hanging stocking is a visual
pun—*bas* is French for "stocking";
courtesy James LeBas

First published in English in 2021 by
The American University in Cairo Press
113 Sharia Kasr el Aini, Cairo, Egypt
One Rockefeller Plaza, 10th Floor, New York, NY 10020
www.aucpress.com

First published in French in 1839 as *L'Obélisque de Luxor. Histoire de sa translation à Paris*.

ISBN 978 1 617 97995 8

Library of Congress Cataloging-in-Publication Data
CIP data applied for

1 2 3 4 5 25 24 23 22 21

Designed by Jon W. Stoy
Printed in China

For William Joy

The Obelisk of Luxor
The story of its journey to Paris

DESCRIPTION

OF THE WORKS THAT IT REQUIRED,

WITH AN APPENDIX

ON THE CALCULATIONS ON THE APPARATUSES USED FOR CONVEYING, EMBARKING, HAULING, AND ERECTING

DETAILS

OF THE PLACES WITH RESPECT TO THE LAND, THE SCIENCE, CUSTOMS, AND TRADITIONS OF ANCIENT AND MODERN EGYPT

FOLLOWED BY

AN EXCERPT OF THE WORK OF FONTANA ON THE MOVING OF THE VATICAN OBELISK

BY MR. A. LEBAS

NAVAL ENGINEER, CURATOR OF THE NAVAL MUSEUM, OFFICER OF THE LEGION OF HONOR

Each people in turn has enlightened the earth,
By the laws, by the arts . . .
(Voltaire, *Mahomet*, Act II, Sc. V)

PARIS

CLARIAN-GOEURY AND VR. DALMONT, EDITORS

PUBLISHED BY THE ROYAL CORPS OF BRIDGES, ROADS, AND MINES

Quai des Augustins, nos. 39 & 41

1839

Contents

Illustrations

Plates

Figures

Translators' Introduction

We will never forget reading the first line of Apollinaire Lebas's account of moving the Paris obelisk: "The brilliant and rapid conquest of Egypt by the French army is inscribed in the annals of history, not only because of the bold military exploits that shook empires, but also because it gave birth to a number of new interests." We both had the same response: Is this guy nuts? Doesn't he know that Bonaparte lost? That the Egyptian campaign was a disaster? That after a horrendous year in Egypt, with his army being decimated by the plague, Bonaparte literally, in the dead of night, deserted his troops and fled to Paris? The French lost! That's why the Rosetta Stone is in the British Museum and not the Louvre. Under the British terms of surrender in 1801, the French had to turn over all the antiquities they had accumulated during the campaign.

When Lebas wrote his opening words, thousands of French veterans of the Egyptian campaign were still alive. Didn't Lebas know of their tales of starvation, disease, and misery? How could we trust the account of a man so divorced from reality? Indeed, the entire introduction is a bit flighty. Lebas ruminates about the nature of the universe, lost civilizations, and presents all kinds of abstract speculations. We wondered whether we actually wanted to translate this book. Amazingly, as we read on, we came to trust the remarkable tale of an even more remarkable man. His blind adoration of Bonaparte was a "one-off," a singular lack of objectivity. We came to appreciate Lebas as a man of great intellect, perseverance, sensitivity, and ingenuity. We quickly realized that his story was a lost classic, something that almost everyone could read with wonder—a tale of adventure full of drama, excitement, and even pathos. It is a story barely known in the western world. Because Lebas's account was originally published in 1839 under the auspices of the Royal Corps of Bridges, Roads, and Mines, it never had a wide distribution. Further, it was published a full three years after Lebas placed the obelisk on its pedestal in the Place de la Concorde. The excitement was long gone, it was old news; Lebas had lost his readers.

We have many reasons for translating Lebas's story. The most important is that it is a wonderful tale that readers will enjoy, but it is also an important piece of history. In the nineteenth century, three obelisks were taken from Egypt and moved to foreign lands. The Paris obelisk was the first. Later, in 1879 and 1881, London and New York would get their obelisks, and those who moved them stood on the shoulders of Apollinaire Lebas. Before Lebas, no one had moved an obelisk since Domenico Fontana moved the Vatican obelisk during the Renaissance. Lebas was a man of modern science and was the first to be able to quantify the great forces that would be involved in moving an obelisk—the friction on the ropes, the strains on the capstans, the distribution of the weights on the pulleys as the 250-ton obelisk was suspended in the air. He was ushering in a new era of scientific engineering, and proudly added an appendix to his story for those who wanted to know the formulae he used in his calculations.

Always on the cutting edge, Lebas intended to raise the obelisk in Paris, not with winches and capstans, but with a remarkable new invention, the steam engine. The engine had been used for years in steamships, but on land its applications had yet to be appreciated. The steam locomotive was still years off and Lebas wanted to show the world what wonders this machine could accomplish. He intended to raise the obelisk with no human power at all, merely using the force of the new invention. When he tried it, he discovered he had been given a defective boiler that could not generate enough pressure to lift the obelisk, so he retooled and went on to Plan B, capstans and man power. Thus, on a cold October day, in front of the king and his family and 250,000 cheering Parisians, the obelisk rose almost effortlessly and came to rest on its pedestal where it stands today.

The scientific aspects of Lebas's lowering, moving, and erecting the obelisk are an important part of our intellectual history, but the human aspect of the adventure is just as compelling. Lebas's sojourn in Egypt was pre-tourism. Very few travelers had been as far south as Lebas. Real tourism would only begin in the 1870s when Thomas Cook & Co. built steamers to take tourists up the Nile to stay at the recently built Luxor Hotel. But all this was far in the future as Apollinaire Lebas struggled to bring Paris her obelisk. When the obelisk was finally safely inside the hold of his ship, the *Luxor*, Lebas and his crew had to wait four months for the Nile to rise before they could sail, so off they went to explore the temples of Nubia. This important, early exploration of Nubia is virtually unknown to the

Lebas's tomb in the Cimetière du Père-Lachaise, Paris; courtesy Olivier Lemaitre

Egyptological community and is certainly worth reading because it adds to our knowledge of early travel in Egypt. However, it is also important because of Lebas's ethnographical observations.

He was interested in everyone and everything, but one encounter seems to have interested him more than others. He witnessed an early striptease act and waxes poetic. What he saw was "The Dance of the Bee," a standard in the repertoire of the local Egyptian dancers. The dancer pretends that she is being plagued by a bee that is under her clothes. One by one the garments come off in an attempt to get rid of the pest. The act ends when the dancer, totally naked, faints on the ground and is quickly covered by her accompanist. Lebas was enthralled, and says so.

Lebas is a keen observer of the human condition and one of his most endearing qualities is his compassion for the plight of the Egyptian peasants. Oppressed by their Turkish masters, the fellaheen were forced to labor for no wages and were beaten if they resisted. Lebas is careful to ensure that his workers are paid a fair wage and given medical aid when the plague strikes them, and he attempts to relieve their suffering whenever possible. His dealings with Muhammad Ali, Egypt's ruler, and with the Turkish officials under him are fascinating.

Lebas was a good man with considerable abilities and it is gratifying to report that he was handsomely rewarded for his hard work on behalf of the French government. When the obelisk was successfully erected in the Place de la Concorde, medals were struck in his honor and he was given a generous pension by his monarch, Louis Philippe. Today Lebas lies in Paris's famous Père Lachaise Cemetery along with other French luminaries. His grave, rarely visited, is topped by a beautiful obelisk. There is an ancient Egyptian maxim: "To say the name of the dead is to make him live again." Our hope is that this book will cause many to say the name of Apollinaire Lebas.

Translators' Acknowledgments

When our colleagues heard the story of how Lebas brought the Luxor obelisk to Paris, the translation practically became a group project. Everyone wanted to help! Professor Richard Auletta of the Foreign Languages Department of Long Island University frequently stopped in to join us in translating, and even compiled a mini-dictionary of nineteenth-century engineering terms. Prof. Dawn Offenbecker, in the Physics Department, also of Long Island University, helped considerably by figuring out how to insert all those equations that Lebas so proudly presents in the appendix.

In Virginia, Dr. Armand Thieblot, an engineer, did his own "quick and dirty" translation and sent it to us, which was often helpful in clarifying an engineering point.

When it came time to reproduce the illustrations, William Joy in Michigan, master of cameras and computers, and a remarkable Egyptophile, enhanced the illustrations and made them a thing of beauty.

And in the last stages, our wonderful team at AUC Press helped give a new life to Apollinaire Lebas. Big thanks to our copyeditor Jody Baboukis, project editor Neil Hewison, who seems to know all languages, and production director Miriam Fahmy.

Thank you all for helping to make Apollinaire Lebas live again.

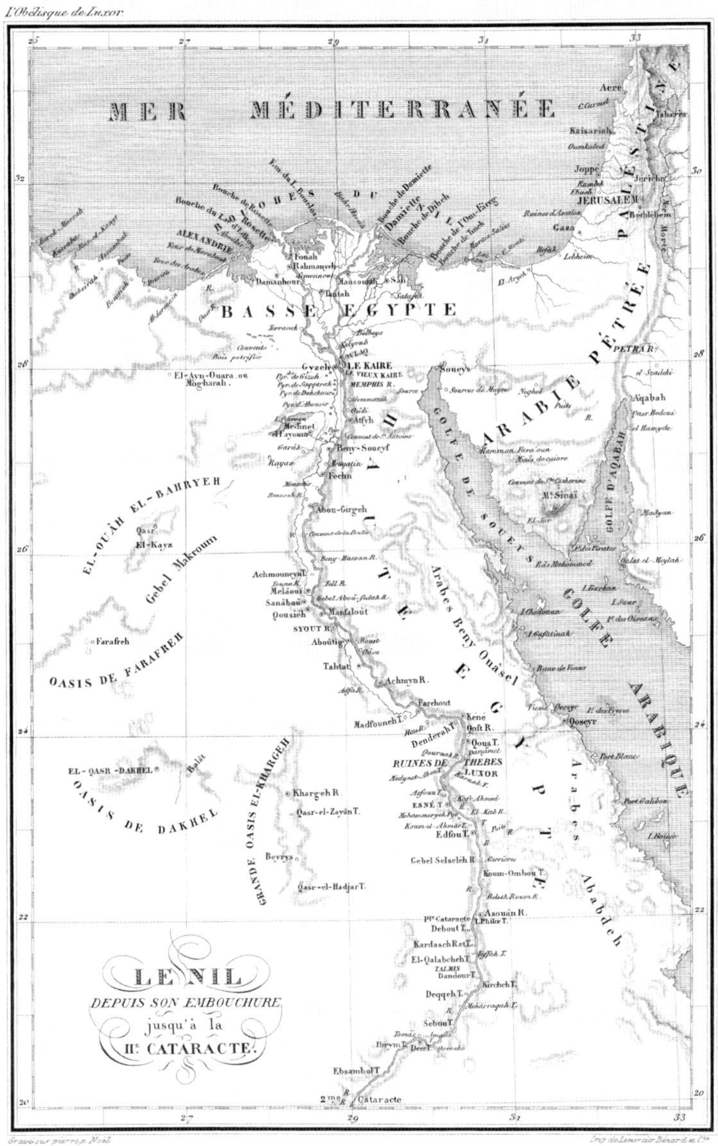

Lebas's Map of Egypt

Introduction

The brilliant and rapid conquest of Egypt by the French army is inscribed in the annals of history not only because of the bold military exploits that shook empires but also because it gave birth to a number of new interests.* If this expedition had produced no other result than to prove the sterile advantage of our military bravery, it would not be anything more today than an interesting episode of the Revolution of 1789. The consequences of the Egyptian campaign do not end there. It provided the means to explore with equal success, to the advantage of Europe's learned, the remains of the most ancient civilization whose memory has reached us. It bears a certain cachet of scientific utility; among the facts produced by the great revolutionary battle, it carries an exceptional and special character among all the events that constitute one of the most beautiful recognitions that France has acquired among nations.

An event whose influence has reflected with such brilliance upon the sciences and arts, which has expanded their vocation and extended their domain, deserved to be symbolized in some way by a monument to be raised within the bosom of the city as a testament of recognition to the heroes and especially the scholars who participated in this event. This monument had to have a local presence, an original character that would remind everyone of the event it was intended to celebrate. There was nothing that could have better reached this objective than the Luxor obelisk. Also, the public embraced the idea of seeing transported to Paris one of these superb monoliths that still remained standing among the ruins of Thebes.

* Translators' note: Lebas seems unaware that Napoleon lost the Egyptian campaign and deserted his men, literally in the middle of the night.

It was left to this generation to achieve by a daring enterprise, of which modern Europe offers no example, the wish expressed two decades ago by the greatest of our sacred orators.* It belonged to the French nation to match the Roman power which, to use expressions from the same orator, enhanced its own grandeur in borrowing the obelisks of the pharaohs from the land of Egypt.

This Egypt, source of all philosophy, ancient cradle of the sciences and arts, no longer exists. How could this nation have disappeared from the face of the world, this nation once so flourishing, so famous for its social institutions, for its profound religious concepts, symbolic emblems of divinized nature! Twenty times invaded by the Persians, by the Greeks, by the Romans, Egypt, after having given laws to the world, was reduced to the state of a Roman province and later inundated by waves of barbarians, which incessantly pushed upon her the ebb and flow of civilization. Little by little, she saw her ancient nationhood lost and extinguished. If some small number of her children escaped the general disaster,† they hardly know themselves today, wandering to places where there are famous cities of which they do not suspect even the past splendor, mixed without any distinction of race or origin with their oppressors. They endure in common with them the yoke imposed by their new masters. Invasion and despotism took everything away from them, even the hope of a better future.

It is mainly from the successive Persian invasions that Egypt has suffered the most as an independent nation. According to Diodorus of Sicily, Cambyses, at the head of a formidable army, covered the two banks of the Nile in blood and ruins. Driven to murder and pillage by this violent and vindictive prince, the Persians ravaged and burned the palaces and temples. The gold, the ivory, the precious stones, the sacred vessels, and all the ornaments that decorated them became prey to his greedy soldiers. Colossi and obelisks were mutilated, thrown down from their bases and cut or smashed to pieces. Cambyses pushed revenge and impiety so far as to violate the asylum of the dead. The mummy of the pharaoh Amasis was whipped on his orders in the public square and was burned in its coffin. That of his queen, robbed of all its jewels, met the same fate.

Twenty-three centuries after these horrible ravages, followed by new conquests and new spoliations, one finds again on Egyptian soil, which seems to be destined to destruction and barbarity, the traces of frenzy of the

*Bossuet. "Now that the name of the king penetrates to the most unknown parts of the world, and that this prince extends far away the research about the most beautiful works of nature and art, would it not be a dignified object of this noble curiosity to discover the beauties that Thebes contains in its desert and enrich our architecture with the inventions of Egypt?"

†These are the Copts, descendants of the ancient Egyptians, who are easy to recognize by their physiognomy, for they resemble in type the ancient Egyptian statues. Their language, it is said, disappeared toward the end of the sixteenth century.

first conquerors. Nevertheless, such was the fertility of the Egyptian art that one still sees standing colossi, obelisks, temples, palaces that were once inhabited by learned and capable priests to whom Herodotus, Solon, Plato, and all the greatest philosophers of antiquity went to draw lessons of science, morality, and wisdom. It seems that those masterpieces of architecture which, by their number and difficulty of construction, seem to accuse us of feebleness and powerlessness, wore out the natural destructiveness of the conquerors. Their gigantic mass preserved them even from the vandalism of modern civilization.

In spite of the numerous accounts that have been published about Egypt, it is impossible for even the most fertile imagination to have an exact idea of the Said (the south) or of Nubia. How, indeed, is it possible to fathom a country where nothing resembles what we have encountered in our childhood, where everything depends upon a unique and annual phenomenon* whose momentary suspension would annihilate all existence? There everything is new for us, the soil, the intensity of the light, the customs, the actions, the characters, the costumes of the inhabitants. There the sky is always clear and serene, the atmosphere always burning. Here torrents of sand cut, cross, and surround the fertile lands with which they seem to be in constant battle. Further, the richest vegetation is never far from the encroachment of the desert. When one sees the silty waters come to fertilize a solitary and arid beach, one would believe oneself transported into another nature. In the midst of so mañy phenomena, in the presence of so many monuments where the power of nature and the mind of man reveal themselves, the mind is inevitably drawn to observe, to study, to look for the solutions to the great physical problems, and mostly to the greatest problem of all, that of social origin.

Who, then, were the Egyptians? Where do they come from? Where do they draw the elements of their science, these people who possessed a complete set of social institutions while Europe was still in the throes of barbarity? Many famous historians devoted their evenings to searching in vain for some authentic testaments about those long centuries of formation and growth. The name of the first civilized people, the part of the globe they occupied, is still a mystery to archaeology. The numerous documents gathered by these scholars only gave rise to a multitude of contradictory systems, which one tries in vain to reconcile or coordinate.

Some, such as the scholar Bailly, bring the origin of the sciences down from the plateau of Tartary. They suppose that a few families escaped to the plateau and so escaped the last world disaster (the flood). Neither accepting nor rejecting this assertion, one can accept that the inhabitants of Africa also found refuge from the inundation on Mount Taranta.† The traveler Bruce

* The periodic overflowing of the Nile, whose waters carry a silt that fertilizes Egypt.

† Mountain situated in Abyssinia near Dixan.

considers this ancient rock as one of the highest mountains of the world and notes that its flanks are pierced with a great number of caves that had probably been used as dwellings for the first inhabitants of the earth. A fertile plain covered today with rich crops surrounds its summit, which is crowned by cedars of all kinds. Vast depressions form several natural cisterns where rainwater collects. A few men were therefore able to retreat there, escaping and thus able to survive a flood that submerged a great number of less elevated places of the globe that were lower than the peak of Taranta.

Several historians who place the cradle of sciences and art in Egypt look upon China and India as countries conquered and civilized by the pharaohs. Huet, who is the bishop of Avaranches, and Mairan believe that Sesostris, at the head of a hundred thousand men, went into China and founded the Chou Dynasty.

The Brahmins, it is true, believe for their part that the Hindus, having gone to China, very slowly abandoned their wild lives and submitted themselves to a single ruler who assembled them into a society, whose antiquity and civilization can be confirmed in books written over a span of more than thirty-three centuries. But their testimony is already hinted at in a passage of Confucius* transmitted to us by Sir William Jones, instructed by the Brahmins, just as Herodotus was by the priests of Memphis.

* "I could, like other writers, report, as simple moral lessons, the facts of the first and second race of our emperors; but I will not speak more, because it does not seem proven to me."

That the invasion of this part of Asia was made by Egyptians, or at least by men coming from the banks of the Red Sea, is further confirmed by the famous de Guines and by the Jesuit Premare, who knew Sanskrit perfectly.

However diverse these opinions might be, no commentator shows us all phases of the Egyptian race's existence until its complete submission to the Roman Empire. It would, however, be profoundly interesting and spectacular to see Thebes born, rise, grow, expand on both banks of the river, and become the most beautiful, the largest, and the most astonishing city in the universe. And to finally follow the successive progress of the great combination of geniuses that could produce so many marvels! But however contradictory these traditions of the past may be, they agree on this point—that the first elements of society come from China or Egypt. If one poses the question of priority, the probability will be in favor of the mother country of Sesostris. No writer so far has put forth the idea that the pharaohs had, for ancestors, the Chinese emperors, while conscientious men versed in oriental languages accept the opposite. This observation tilts the scale to the side of the Egyptians, who have always been considered correctly as a seminal race and type. Egypt developed everything on its own, without borrowing anything from other regions. Is there one on earth that bears the cachet of great antiquity more than Egypt? Its great antiquity is proven, without question, by the zodiacs sculpted in the tombs and on the ceilings in sacred buildings. The state of the sky as presented to us by the Zodiac of Dendara is already no longer the same as the one that sculptors later drew in the necropolis of Thebes. By following the precession of the equinoxes, this last one is no longer the sky that we see today. A thousand other testaments add themselves to this immense antiquity and a profound knowledge of astronomy. There are the Jews coming out of the bosom of Egypt, today scattered on the whole surface of the earth. There is the art of the sculptor on every piece of stone, art that is not essential to the lives of people, and in which the Egyptians excelled two thousand years before our era. There are the gigantic constructions with which the ground still bristles, and which show us the richest museum in the universe. These constructions, the least ancient of which existed ten thousand years before Plato* and which we may still come to admire when there are no more vestiges of our sumptuous cities, are built partly with the rubble of other constructions, collapsed with age. This indicates two eras of monuments, two ages of architecture.

The time of human existence, measured in this way, unfolds in the imagination in an indefinite immensity. The impossibility of going further back in time to enumerate the previous ages which are proven by material facts, the impossibility of assigning a certain origin to these ages whose boundaries are pushed back by new discoveries every day, this impossibility of doing all these things, would it not suggest to some philosophers that the idea of physical eternity, such as the feeling so deeply ingrained in us of God's existence, has created in us the idea of moral eternity?

The Greeks, who owe their arts, their sciences, their philosophy, their laws, and even their gods to the Egyptians, do not really teach us anything positive on the true state of Egypt until the time when they penetrate it. As for the Romans, they explored Egypt as conquerors rather than conservers. All they did was force Egypt to submit to their military power to get the grain they would distribute to the people. They left her in the hands of a prefect. The lovers of antiquity must regret that these two peoples neglected to preserve the hieroglyphic alphabet, which they probably knew, since several buildings raised during their rule are covered with hieroglyphs completely similar to the ancient ones and which represent the cartouches of Ptolemy, Cleopatra, etc.

Holy Scripture praises Moses for having been instructed in Egyptian wisdom. However, he keeps absolutely silent on the customs, laws, and religion of the ancient inhabitants of this country. Uniquely occupied by his divine mission, he reports hardly anything

* Plato, second book of *Laws*. "If one wants to pay attention one will find in Egypt works of painting and sculpture which are no less beautiful than those done today and were done precisely according to the same rules ten thousand years before us. When I say ten thousand years, this is not an expression, it is an exact truth."

other than the facts directly related to it, but from these facts we can deduce that more than twenty centuries before the Christian era, Egypt was ruled by a wise, orderly government based on respect for laws and customs which existed from time immemorial; that long before Cecrops appeared in Attica,* religion, ethics, politics, knowledge of astronomy, the most useful arts of society, and even the entertainment arts had reached a point of perfection that required several thousand years of culture. We see the political prophet and legislator [Moses] utilizing in the desert the knowledge drawn from the sacred schools. He melted down the golden calf. Helped by the Israelite workers from the workplaces of Thebes and Memphis, he had to make all the necessary objects for worship which, according to the details that he himself gives of them, needed the participation of a great number of artists, sculptors, blacksmiths, goldsmiths, carvers, etc. And that is not all. Some authors have asserted that Moses, initiated into the most secret mysteries of the priesthood, probably used gunpowder to present the burning bush, to split the rock, and to cause Korah, Dathan, and Abiram [leaders of a revolt against Moses] to be swallowed by the earth.

Herodotus, who voyaged in the Said after Moses, is undoubtedly worth believing when he recounts his own observations and relates what existed in his time, but we see he also merely gathered from past centuries vague notions, often childish stories. This philosopher merely saw in passing the priests of Memphis and Thebes, and he only consulted with them via the mediation of a translator. How, out of such elements, could he have been able to compose the veritable history of the Egyptian people, of their language, their deeds, and their men? Therefore, he was contradicted by Manetho and accused of lying and inventing.

Manetho, an Egyptian of the priestly class, wrote in Greek the universal history of his country, which he claimed to have taken from the sacred archives of the temple of Heliopolis, where he had the function of sacrificer. Unfortunately, of this work, known under the name of *Chronicle of Manetho,* we only possess incomplete and contradictory fragments that were preserved for us by Josephus, Julius Africanus, and Eusebius.

The accounts of other historians do not deserve much more trust than those of Herodotus when it concerns the origins of the Egyptians or their writing, or the principles of their primitive constitution, their history, the development and progress of their arts. Facts and dates all are uncertain, vague, mixed with obvious fables. There is contradiction in the length of the measurements, in the names and the geographical positions of the cities and remarkable sites of this astonishing land. As for modern writers, they only reproduced or commented on classical authors without being able, until now, to bring light into the depth of

* Translators' note: Cecrops was the first king of Attica.

this chaos. From the historical point of view, the Egypt that the centuries have given us resembles a colossal machine whose principal organs were annihilated forever, dispersed on the ground and partially covered by the sands. The debris of this machine, although poorly studied, poorly understood, nevertheless announces a greatness of conception, a boldness, an originality of execution which can only be the result of an immense superiority in science and ability of crafts. We can see clearly all the greatness that this powerful mechanism has been able to produce. Until now nobody had managed to show us the genius which made possible her construction, to explain the workings of levers and hidden springs which made the whole thing move with such perfect harmony, to calculate its power, its effects, to retrace its primitive form and its successive improvements up to the moment of its destruction. To leave this mathematical language, we will say that there exist neither translations nor excerpts of Egyptian books dealing with philosophy, industry, and morals. It is not that this nation, both moral and glorious, in her past, has neglected to transmit her history to posterity. The ancient tradition is there, right under our eyes, engraved on the hardest stone that the earth contains. The public and private life of the Egyptian, in all the situations where man can find himself between childbirth and death, are laid out for us to see. Their successive discoveries, their methods, are sculpted on the sides of obelisks, on the walls of edifices, in the tombs of kings and private people with characteristic signs taken from nature, which all together form diverse scenes of a perfection which the modern artist could never surpass. The genius of Egypt has foreseen everything. It has calculated everything for a vast future. The hieroglyphs still exist in all their purity, shining with their most brilliant colors. But it was in the decrees of Providence, who does not care about man's plans, that the first people on earth would one day be brought back to the lowest step on the social ladder; that one would wonder, on the very site of Thebes, what language Sesostris himself spoke; that one would look for and that one will always look, perhaps always in vain, for the art to decipher characters that alone could reveal the history of those faraway ancient times. This question [the translation of hieroglyphs], which is of the highest degree of interest to philosophy, science, and the arts, which has occupied scholars for so many years, has remained up to the present without a complete solution.

It was reserved, however, to a man of vast knowledge, Champollion the Younger,* to carry the flame in this deep obscurity and to resolve this great problem. He had already lifted the corner of the veil that covers the sacred and profane history of Egypt, so as to reveal a new past to our eyes, when death caught this illustrious

* Translators' note: Jean-François Champollion had an older brother, Jacques-Joseph, who also worked in the field of Egyptology. To distinguish them, the decipherer of hieroglyphs is called Champollion the Younger.

Orientalist in the middle of his work. May the honors rendered to his ashes create a genius equal to his, who, more fortunate, may penetrate to the bottom of this mysterious quarry which our countryman and scholar has opened!

The reader must not look on these preliminary observations as the summary of a work meant to develop them. This is not the mission of the author. The text that he publishes has mainly as its goal to present accounts of facts in all their simplicity, the series of facts relating to the lowering, the shipping, and the erecting of the obelisk of Luxor. If sometimes he strays from his topic, if for example several weeks of leisure allowed him to explore the valley of Egypt and part of Nubia, what he saw he has recounted faithfully, without other pretension than to fill the amount of time till the completion of the work up to the launching of the *Luxor*, with some details of the monuments, the products and cultivation of the soil, the manners and customs of the inhabitants of these two countries! In order for the account to be clearer and more precise, this book is divided into several distinct parts.

The *First* concerns the departure and crossing of the vessel *Luxor* and the lowering of the monument.

The *Second* is devoted to a quick excursion made in Nubia and to some observations of its customs, and to the return to France.

The *Third* deals with the work in Paris, that is to say the unloading, the transport, and the erection of the monument on the Place de la Concorde.

The *Fourth* gives the translation of the account of the former Italian engineer, Fontana, who in 1586 erected in Rome the obelisk of Caligula. This account, from which only the parts irrelevant to the question of art have been removed, allows us to compare the means used, two and a half centuries apart, to move great masses.

To these four parts we add an appendix where, independently from some considerations of the mechanical knowledge of the ancients, we present the calculations which determined the limits of forces and resistances of the system and determined the choice of where they should be applied in the various maneuvers of which the obelisk of Luxor was the object.

Part 1

Departure — Crossing of the Vessel *Luxor* — Lowering of the Monument

Chapter 1

First negotiations about one of the Egyptian obelisks — The opinion of Mr. Champollion the Younger on the obelisks of Luxor — The mission of Mr. Taylor in Egypt — The giving to France of one of the obelisks of Luxor by the viceroy — Construction of the vessel *Luxor* — Preparations for departure

It is said that Napoleon first had the idea to transport to France one of the Alexandrian obelisks. Even if the great man did not execute this project, he must have conceived of it.

After having brought to the land of the pharaohs our victorious arms, the conqueror of Egypt, called back to France by events that his presence there could have prevented, he was indeed honor bound to give to his country such a trophy, and leave after him an irrefutable witness of our astonishing campaign of 1799.[*] This same sentiment had already given Rome the possession of obelisks that decorate this ancient capital of the world, immortal monuments of its victories and sole remains of its conquests.

Any time one thinks of monumental magnificence, one thinks of obelisks. The pharaohs, the Lagids,[†] the Caesars, the popes have one after the other erected at great cost these imposing symbols of Egyptian civilization. Thebes, Alexandria, Byzantium, Rome had theirs. Paris had to have its own, and it was Thebes that provided it.

Transported 1,800 leagues from the land where it was erected for the first time thirty-three centuries ago, the obelisk of Luxor, henceforth saved from all events that could threaten its conservation, stands now on the largest, most beautiful of our squares, where it brings back glorious memories. The population of Paris followed intently the smallest details of the unloading and the placement of this monolith. They visited with a curious solicitude the vessel that transported it. They were touched at the

* Translators' note: Although Lebas uses the year 1799, the Egyptian Campaign actually began in 1798 and ran through 1801.

† Translators' note: The Ptolemies.

thought of the work, the exhaustion, and the real dangers that had to be surmounted to uproot this admirable monument from the sands of Thebes, and to bring it all the way to the banks of the Seine. The part played by each in this project should rightly be known today.

It is important to tell how the result has been attained.

We have spoken of the idea attributed to Napoleon. What twenty years of war and the continental blockade did not allow him to undertake, Louis XVIII began. It was on his orders that the consul general of France in Alexandria was asked to negotiate with the viceroy of Egypt the ceding of one of these monuments. The negotiations on this subject had a great success. The viceroy gave one of the two so-called needles of Cleopatra. The second was given to England.

It would be difficult to tell precisely why the obelisk that was presented to France was not brought back. The fact is that it remains on its base in Alexandria where it is today.*

Sometime later, the love of knowledge brought to Egypt Mr. de la Borde and later Champollion the Younger. It was those two scholars who, struck by the beauty of the monuments of Egypt, reawakened the memory of the gift to Louis XVIII.

Mr. d'Haussez was then minister of the navy. Baron Taylor's rank, and even more his studies, put him in a position to enlighten this minister. He reminded him that there existed in Alexandria an obelisk, belonging to France, which might slip from her hands; if we neglected to have it removed, it would give our rivals the pretext and time to appropriate it.

The Egyptian Museum to which King Charles X has just recently given his name had been formed. People were working to enrich it with all the fragments of Egyptian antiquity that could be obtained. Would this not be the moment to guarantee the possession of a monolith which alone could add so much value to this interesting collection?

So stated, the question was soon resolved. Baron d'Haussez adopted with enthusiasm the ideas of Mr. Taylor and was determined from that moment on to leave a legacy of himself in the ministry by mounting an expedition that he was to preside over and that he considered would give to himself and to the reign under which he lived a claim to glory.

On November 18, 1829, this minister wrote to Mr. de Cerisy, engineer of the French Navy, then in charge of directing the naval constructions of the viceroy of Egypt. He wrote to ask him the dimensions and weight of the obelisk, as well as his opinion regarding the means to take it down and transport it. An officer of the navy employed in the French fleet in the Levant was at the same time given the job of discussing the subject with Mr. de Cerisy.

However, Mr. Champollion the Younger, touring Egypt, had seen the obelisks of Thebes. Struck by the beauty of

* It must be said that, even though the second monument had already fallen down, England has failed to obtain it, and the plans made for its transport have still not been carried out.

these monuments he expressed himself thus in a letter that was later communicated to the minister of the navy:

> I am very glad that the knowledgeable British engineer had the good idea to discourage his government, and all others, from building a road costing 300,000 francs for these poor Alexandria obelisks. They are so insignificant when compared with those of Luxor. If we are going to see an Egyptian obelisk in Paris, it should be one of those at Luxor; there will remain the one at Karnak, the most handsome of all, but I will never give my support (of which there is no need) to a plan to saw in two or three parts one of these magnificent monoliths; this would be sacrilege!

Such an observation coming from a man as competent as Champollion the Younger had to make an impression on Baron d'Haussez. So, this minister did not hesitate to give all preference to the Thebes obelisks. In his opinion, it was much less important to transport to Paris the Alexandrian obelisk than to obtain the ones of Thebes, even if one had to give up Cleopatra's Needle to get them.

It was according to these new ideas that plans of operations presented by several people were conceived, particularly a project drawn up by Mr. Besson, officer of the French Navy in the service of the pasha of Egypt, a project for which the first deliberations took place in Paris, concerning the measures to take to remove and transport the obelisks.

Baron Tupinier, director of ports at the Ministry of the Navy, had suggested from the beginning that one could load those heavy masses into the hold of one of the 900-ton transport ships belonging to the state by temporarily removing either the bridge or the front of the ship to get them on board.

Mr. Besson, on the contrary, remembering the means formerly used by the Romans for such operations, suggested building in Caramania a raft 33 meters long by 13.35 meters wide, and having it towed by a merchant ship to Damietta or Rosetta and from there up to Luxor or Karnak by *canjiehs*.*

According to his project, the obelisk to be transported had to be enclosed in a cylinder of pinewood strongly reinforced by iron bands and then placed in the middle of the raft lengthwise in such a way as to jut out of the top of the surface by half its diameter.

This means of transportation could only be used to take the obelisk with the help of a steam tugboat as far as La Seyne, in the harbor of Toulon. There a ship would have been built expressly to bring it to Le Havre, from where it would have come to Paris by means of ships used for navigation of the River Seine. This project, of which we only give a very brief summary, was, as one can see, very complicated.

* See * footnote page 27.

The flaws its execution would have presented, the errors the author made in the dimensions of the raft, the difficulties that made the operation impractical, all these did not escape the careful examination of Baron Tupinier, who in a report submitted to the minister on November 12, 1829, proved, in a most obvious way, the need for other ideas.

Baron d'Haussez approved the conclusion of this report and decided that it would be communicated to a commission consisting of Messrs. de la Borde, Tupinier, Drovetti, Taylor, Briet, de Mackau, and de Livron, of which he reserved for himself the presidency. This commission, in fact, convened on November 19 in the cabinet of the minister.

After lengthy discussions about the best way to transport the obelisks of Thebes to France, the commission concluded, following the proposal of Baron Tupinier, that Baron Taylor should be sent to Alexandria to discuss with Messrs. de Cerisy and Besson the best plan to adopt, and that at the same time one of the ships *Rhinoceros* and *Dromadaire* would be held ready to be sent to Egypt to get the obelisks, if it was agreed that it was possible to load them onto these ships.

However, it was first necessary to get the viceroy of Egypt to cede the obelisks. This was where the mission of Baron Taylor had to start. The known dispositions of Muhammad Ali toward France did not leave any doubt as to the success of the steps to take. So Baron d'Haussez did not hesitate to submit to the king the proposal regarding this affair. We believe that we should transcribe literally the report he presented on this subject to Charles X, who approved the conclusions on November 25, 1829.

Paris, 25 November 1829

Sire,

France owes to her kings the most beautiful monuments that decorate both her and Paris, which, with the exception of only one capital of modern Europe, will soon contend for the foremost position with the most famous cities of ancient times. But we must admit that Paris, with its palaces and public squares, has not yet reached the degree of splendor of Rome, although the capital of your kingdom rivals Rome in magnificence. One sees in Paris none of these obelisks transported from Egypt to Europe. Your august brother who, like Your Majesty, accorded to the arts such an enlightened protection, had ordered to negotiate with the Pasha of Egypt to obtain the Alexandrian obelisks, or others which are found in this ancient land, richest in ancient artifacts. Muhammad Ali, eager to respond to the wishes of the King of France, offered one of the Alexandria obelisks called *Cleopatra's Needle*. Unfortunately this obelisk, on which time has exerted its ravages, presents less interest for the arts than those that are still standing in Luxor, in Upper Egypt. Admirable as they are because of their precious workmanship according to archaeological reports, if Paris possessed these, it would no longer envy Rome, and their erection in public places, besides making them more beautiful, would be a reason of deep gratitude on the part of the hardworking and educated men who devoted their lives to the study of antiquity.

The good disposition of the Viceroy of Egypt toward France, the possibility of transporting these monuments on one of the boats of your navy without great cost, the powerful desire to complete in a way the collection of Egyptian monuments for which such sacrifices have been made, the zeal that ought to be shown to finally gather the fruit of former negotiations, and not be deprived of the monument already obtained, all prompt me to propose to Your Majesty to charge Baron Taylor, known in the arts for his work, for several voyages in Africa and Asia, and known for his service to the King, with a particular mission for this object to the Pasha of Egypt. The purpose of this mission would be to negotiate the exchange of the Alexandrian obelisk for the Luxor obelisks, or to obtain the obelisks of Thebes without giving up that of Alexandria.

It is with the conviction that the result of this operation will only add to the éclat of the reign of Your Majesty that I beg you to approve the requests that I have just had the honor to submit to Your Majesty.

If Your Majesty deigns to approve this project, I will gather around me a special commission composed of Alexandre de la Borde, de Livron, Drovetti, de Mackau, Tupinier, and Taylor, who, before the departure of the latter [Taylor], would conclude the negotiations and would combine its means in a manner to ensure success.

Signed Baron d'Haussez

Once this report had received the approval of the king, the only concern was preparing the means of execution. It was then that, little satisfied by the answers that came to him from Alexandria about the plan to transport the obelisks on one of the ships that was in Toulon, the minister of the navy charged Baron Rolland, inspector general of maritime engineering, with the purpose of devising a definitive plan of operation.

Baron Rolland proposed to build in Toulon a ship designed to go up the Nile and receive in its hull one of the Thebes obelisks, and then bring it to Paris.

Baron d'Haussez, convinced that this was the best plan of all of the former proposals, did not even wait for the departure of Baron Taylor, nor for the official gift of the obelisks, to order the construction of the ship that, before beginning construction, was named *Luxor*.

The ship *Luxor* was destined by the nature of its mission to navigate on the sea and two rivers. So the problem to resolve consisted first in giving it dimensions and form that would allow it to pass under the arches of the bridges of the Seine and that, with the obelisk on board, it would not submerge more than two meters in the water, in accordance with the river navigational laws. In order to reach this goal, Mr. Rolland was forced to stray from the usual rules of naval architecture. The length of the *Luxor* was not proportional to its width, and the surface of the hull was very close to a parallelepiped whose angles and edges were blunt. Its ribs were composed partially of pine, and the hull reinforcement in copper was substituted by wood planks to lessen as much as possible the cost and weight of the ship.

It was not enough to just coordinate these data that were contrary to the functioning and safety of a ship. New conditions demanded other modifications no less important.

The *Luxor* had to land on a sandy beach in a perfectly straight position without the help of stanchions or stays, and to carry a monolith whose mass was out of proportion to the space it would occupy in the hold. It was these conditions that this engineer had to combine with the ship's frame. Thus he had to fit the ship with five keels in order to create supports on each side of the main keel and by this means divide the forces exerted by the cargo to a greater number of points. He had to reinforce the walls by oblique struts and he had to increase the number of the longitudinal ties in order to integrate the system. These difficulties, in addition to all the others, are worth mentioning. In summary, Mr. Rolland did his best, and the experience justified his decisions.*

However, Mr. Taylor, busy getting ready for his departure, was nominated commissioner of the king to the pasha of Egypt by royal decree on January 6, 1830, to negotiate the grant of the Thebes obelisks and to have Cleopatra's Needle transported to France. He was also charged to obtain in Egypt objects of art and antiquity destined to enrich the Musée Royal du Louvre.

He was given several presents designated as gifts for Muhammad Ali and his son. A credit of 100,000 francs was opened to take care of the expenses of his mission. And a brig, the *Lancier*, was ordered to bring him to Alexandria.

Mr. Mimaut, consul general of France in this city, himself an enlightened lover of the arts, had been told

* The recent death of Mr. Rolland is a subject of regrets as deep as they are deserved.

of the object of Mr. Taylor's mission and was invited to help with his influence and esteem. Mr. de Cerisy had also received the same alert, so there would be no doubt about their useful cooperation.

The *Lancier*, delayed by bad weather and contrary winds, did not arrive in Alexandria until April 23, 1830. Mr. Mimaut was not there then. He was in Cairo with the viceroy, but upon leaving he had left instructions with Mr. Bottu who was replacing him in his capacity of vice-consul. It was this latter who received Baron Taylor and introduced him to the son of Muhammad Ali, Ibrahim Pasha, who was then in command of Alexandria in the absence of his father. After an excellent welcome by this prince, Mr. Taylor departed for Cairo, where he did not find the pasha, and it was only on his return to Alexandria at the end of May that he was finally able to meet him.

However, Mr. Mimaut, as soon as he first received the information of Mr. Taylor's mission, had immediately told the pasha about it and had already disposed him to favorably welcome the request that the commissioner of the king was charged to present to him. Mr. Taylor, who has never ceased praising the cooperation of Mr. Mimaut, must have been aware of this help when on May 31 of the same year he had an audience during which they both asked and received the grant of the three obelisks which today belong to France.

This interesting negotiation was, however, not without its difficulties, and it took all of Mr. Taylor's skills, talent, and perseverance to surmount the first obstacle that he had to overcome. By neglecting to have the obelisk of Alexandria removed, it seemed that France had somehow lost ownership of it. On the other hand, the pasha had given the obelisks of Luxor to Mr. Barker, the consul of England, who had vigorously requested them.

But the desire to please the king of France, to whom, as he said, he wanted to refuse nothing, and also, without a doubt, the skillful intervention of Mr. Taylor, suggested to the viceroy the means to conciliate everything. He gave to France the two obelisks of Luxor, confirmed the previous gift of the Alexandrian obelisk, and offered in exchange the Karnak obelisk to Mr. Barker. The latter accepted the proposed exchange; thus this business was entirely concluded. It was agreed, to avoid any later difficulty, that before returning to France Mr. Taylor would receive an official document attesting to the gift of the three obelisks.*

---

* Translators' note: Although the French had been given three obelisks—the two at Luxor Temple and one of the two at Alexandria—they only took one. Before Lebas sailed for Egypt, the French sent a ship, the *Dromadaire*, to Alexandria to bring home one of the two there. In their hurry to retrieve an obelisk, the French had not realized that timber was not readily available in Egypt and there was not enough wood for the scaffolding needed to take down a standing obelisk. Consequently, the *Dromadaire* returned to France with its hold empty. Several years later, when the French saw all the difficulties that Lebas had had in transporting the one obelisk from Luxor Temple to Paris, all

Meanwhile, in Toulon, there was a rush to build the *Luxor*. Already everything was ready in France for the expedition. Mr. Mimerel, engineer of the navy, had at first been designated to be in charge of the work of taking down and loading the obelisks, and of the works necessary for their erection in Paris. On the other hand, Mr. Verninac de Saint-Maur was charged with command of the ship.

The *Luxor* was launched on July 26, 1830. On the 27th the order to expedite the commission of the ship by all possible means was addressed to the naval prefect of Toulon. Already the *Dromadaire* had been sent to Alexandria, where it had arrived by June 25.

Mr. Taylor, who was then traveling through Egypt with the purpose of gathering several objects of art for the museums of France, learned through letters brought by the *Dromadaire* of the progress of the construction of the *Luxor* and the plan to send it around the month of September. He thought that he had to not only load the Alexandria obelisk on the *Dromadaire*, but also, by the same means at his disposal, take the second obelisk of Luxor, and, according to the indications which Mr. de Cerisy would provide, transport the second obelisk of Luxor in such a way that these three monoliths would arrive in France towards the end of fall. Such a result, we must say, was impossible. Flattering himself that he could achieve this goal, Mr. Taylor had consulted only his own zeal, without foreseeing the difficulties that would stop him right from the beginning.

Such was the state of things when the viceroy of Egypt learned in Alexandria the events of July 1830 [the revolution]. The details had reached him through letters of commerce that only gave the news up till August 4. Fanned by the uncertainty born from this news, intrigues to take away from France the admirable monuments that had become her property started up again. The English consul conceived a plan to have them removed, and several agents of the pasha agreed with him. The rumors even spread that the viceroy did not want to let Cleopatra's Needle embark on the *Dromadaire*, in exchange for which it was said that he had given the two obelisks of Luxor. But this rumor, born in the antechambers of the harem, had then sprung to the cafés of Alexandria,* and from there all the way to Paris. This rumor did not rest on anything true. It was just the talk of idle or jealous people, always ready to antagonize or deprecate the most noble enterprises.

Besides, Baron Taylor and Mr. Mimaut had not given France's rivals any time to express their new desires to Muhammad Ali. Although they did not

plans to retrieve the other two were quietly forgotten. Eventually, the two obelisks at Alexandria known as Cleopatra's Needles were given to the United States and Britain and erected in Central Park and on the Thames Embankment respectively. The Hatshepsut obelisk at Karnak originally promised to Britain remains in situ, as does the second obelisk at Luxor, which until the late twentieth century was still legally France's to take: only in 1984 did President Mitterand formally give it back to Egypt.

* A few Europeans had established cafés in Egypt, imitating ours.

know the decisions that would be taken with regard to the transport of the three monoliths, they had perfectly well understood that in any case the unquestionable acquisitions of France had to be preserved.

It was not hard to persuade the pasha that the July revolution should not be a reason to rob France of a gift that he had made to the country, that such monuments, even in his thoughts, could not have been given to the person of the king, but to the nation itself, whose irrevocable property they had become. They finally succeeded in obtaining from the pasha the full and complete confirmation of his first disposition. The gift of the obelisks to the government of Louis Philippe was sanctioned in a letter that this sovereign had his minister write officially on November 29, 1830 to Count Sebastiani, then minister of the navy.

This important document must find its place here in its entirety.

Alexandria, November 29, 1830

Your Excellency,

His highness, the Viceroy of Egypt, received from Baron Taylor the dispatch he was carrying from the Minister, Secretary of State of the Navy and its colonies, to negotiate in the name of His Majesty the King of France, and to obtain one of Cleopatra's Needles at Alexandria, and in particular the two obelisks at Luxor which are part of the ruins of Thebes.

His Highness the Viceroy asked me to express to Your Excellency the satisfaction that he feels in showing his gratitude to France for the numerous marks of kindness and friendship which were shown to him at different times and which have been recently renewed by His Majesty King of the French, through the Consul General Mimaut.

I have been ordered by His Majesty to put the three above-mentioned monuments at the disposal of His Majesty the King of the French as of now, and we pray Your Excellency to please offer respects to His Majesty in the name of His Highness the Viceroy Muhammad Ali Pasha.

It is very flattering for me to be the interpreter of the desires of my prince on this occasion and I pray Your Excellency to accept the assurance of my very distinguished consideration.

Signed, Boghoz Joussouf

We have previously seen that the *Dromadaire* had arrived in Alexandria on June 25, in the absence of Mr. Taylor. It had with it the apparatus believed to be necessary for the loading of the obelisk. This ship, however, did not bring from Toulon all the wood necessary to accomplish this project. It was necessary to buy it in Caramania, for the stores of Alexandria did not possess a single piece of wood that the pasha could give to the project. The delays arising from this circumstance were the cause which later would result in giving up on the removal of this monument.

However, the change that occurred in the government of France did not make the minister of the navy lose sight of the business of the obelisk. The leader that France had just chosen for herself was himself too much a friend of the arts, and above all, he was too jealous of our national glory, to abandon under his reign an enterprise designed to increase the country's luster. Thus they continued to be occupied with the commission of the *Luxor*. The instructions prepared for Messrs. Mimerel and Verniac de St.-Maur had been sent on September 22, and a telegram on October 6 had ordered them to hasten by all possible means the departure of the ship. On his side, Mr. Champollion, who was then in Paris, had by no means forgotten a project whose execution was one of his main thoughts. He made sure it remained an important consideration. He said in a note of September 29:

> The former government had no other intention than to embellish the capital with a new decoration. The national government has to have a higher and greater purpose. It alone should mark its first steps by eternalizing the glorious memories of triumphs of our army during the wars of the republic. These monuments, noble patrimony of the generation that was the first to fight for French freedom, are still awaiting their consecration. Foreigners are surprised to go through our capital without finding somewhere even one monument that recalls, even indirectly, our astonishing campaign of 1799. No monument can better achieve this purpose than one of the obelisks of Egypt.

His conclusion was that they should hurry to finish the enterprise already begun, but he insisted on focusing on the obelisks of Luxor, much more precious, he added, than those of Alexandria. He insisted that between the two, one should give preference to the one on the right, or west, whose inscriptions recall the name of the great Sesostris.

Count Sebastiani had perfectly appreciated what was right in these observations. The *Luxor* was on the verge of departure when Mr. Taylor, who had arrived in Toulon on the brig *Lancier*, requested that the expedition be postponed for at least several days. He had to give to Mr. Mimerel the necessary explanations on the resources that this engineer had to acquire. Besides, the season having passed for navigation of the Nile, they would have to wait for the beginning of the inundation of 1831, which would only occur in July. According to Mr. Taylor, all one could do to lose neither time nor money was to transport to France the Alexandria obelisk and to then return to Egypt in the season favorable for going up the Nile. There was yet another reason to support this opinion: according to Champollion, Cleopatra's Needle was very inferior to the other obelisks, and since no one knew if it would be possible to transport all three of them, it was necessary to think first of the most beautiful of the ones at Thebes. Therefore the departure of the *Luxor* had to be delayed till the following spring.

All the expenses relating to the expedition had been posted to the account of the navy, in accordance with an order of January 6, 1830. The government had granted a credit of 300,000 francs.

The money already spent both for the building and commissioning of the vessel and for the mission of Mr. Taylor left only about 37,000 francs available. Besides, no new credit had been open to the ministry of the navy during the year 1831. This observation naturally raised the question of whether to persist in an enterprise for which at that point the total expenses were not known, or whether, postponing all future operations, it was not better to first send to Thebes an engineer who would give an estimate for the work to be done.

This question was quite serious at this time, when France had to provide for important needs of commerce and industry, and moreover maintain her independence, which she had reason to believe was threatened by continental Europe.

Count Sebastiani did not ignore this difficulty, but seeing the problem and thinking further, he saw how important it was not to leave uncompleted a great and useful enterprise. He also thought, and rightly so, that sending an engineer to give an estimate would mean losing time and money. Besides, he understood the bad effect that would be produced in Egypt by the sudden abandonment of those monoliths so ardently sought till then.

From all those considerations, he concluded that even if it cost 200,000 francs for the transportation of these obelisks, this new sacrifice was better than losing the fruits of the money already expended. In any case, he would have rightly deserved blame if, before giving up this project, he did not at least submit the question to a legislative vote.

The government allotted the credit requested and so the last of the obstacles was removed, which had successively been on the verge of holding up this interesting enterprise. As early as January 19, 1831, the minister of the navy, according to the intentions of the king, ordered the prefect of the navy at Toulon to have everything ready so the *Luxor* could at least be sent at the beginning of the spring.

Mr. Mimerel had hinted that the state of his health would probably hinder the accomplishment of his mission. This prediction was confirmed and it became necessary to choose a successor for him. Consulted on the new choice, the maritime prefect at Toulon proposed me as well as Mr. Moissard, naval engineer.

Designated by the minister to head such an important mission, I accepted with pride such a mark of confidence. My first care was to choose, according to the authorization I had received, the men who were to help me, and the materials I judged necessary. Everything being so arranged, I embarked on the *Luxor*, which was under the command of Mr. Verniac de Saint-Maur.

Here are the instructions that were given to me by Count Sebastiani.

> Sir:
>
> You have received the order to go to Egypt to head an important operation to bring to Paris the two obelisks of Thebes which the Viceroy Muhammad has given to France.
>
> You know that the vessel *Luxor* was built in Toulon for the transport of these monuments and that the water displacement of this vessel had been calculated in such a way as to allow it to sail up the Nile near the place where they are.
>
> I gave orders for the *Luxor* to be ready to leave as soon as the beginning of this month. You will embark on this vessel as a passenger. It will bring you to Alexandria and there you will decide if it is better to stay on board till Thebes or if it would be better to arrive in Thebes before it does.
>
> You will examine with greatest attention the location of the two obelisks and the condition of their surroundings to establish a work plan that you will implement to lower the monuments, move them to the banks of the Nile, and put them in the boat that is to bring them back to France.
>
> If, as I presume, only one obelisk will fit on the *Luxor*, the operation will be made twice; it will be necessary to begin with the one on the right, which was said to be the more precious, as you will see from the copy here appended of a letter by Mr. Champollion the Younger, who has recently copied and translated the inscriptions.
>
> You will be specifically and exclusively charged to direct all work relative to lowering, transporting on land, and loading each of the obelisks, as well as to direct all those who will

be needed to prepare the land for the arrival of the monument on the bank of the Nile.

You will work with the officer who will command the *Luxor* on the choice of where you will anchor it, and on all the methods employed to load the obelisk; you will decide whether to run the ship aground, remove its bridge, or open one of the extremities, and this officer will be your second in all aspects of the execution of these various works. You will also consult with him so that the resource materials on board will be applied as needed.

As soon as the obelisk is loaded and you have replaced the parts of the boat that you needed to demolish or remove, the *Luxor* will be the responsibility of its captain, who from this moment will have sole responsibility to conduct it to Paris; your task will then end, until such time as it becomes necessary to recommence the same operations for embarking the second monument.

Perhaps it will be possible to employ other methods to transport one of the obelisks. For this purpose, Mr. Mimaut, France's consul general to Egypt, will place at your disposal Mr. Linant, a Frenchman long established in the country and who seems very intelligent.

I append here a copy of a letter in which Baron Taylor furnishes various information on the subject that, without having anything really positive, might nevertheless be at least good to consider, especially concerning the choice of objects to take from Toulon.

I am giving to the maritime prefect of Toulon the order to put at your disposal, through various directions of this port, all materials that you will find necessary for the execution of the operations mentioned.

If when you are in place you find other resources indispensable, you will probably find them at Alexandria through the intervention of the consul general of France, and if it is impossible to obtain them in Alexandria, you will write a request to Toulon.

You will choose in the arsenal of this port a master repair carpenter, a second to the master of the same profession, and six workers who will be carpenters, drillers, caulkers, and if needed joiners; you will add a blacksmith, two good metalworkers, and two stonemasons.

All these masters, sub-masters, and craftsmen will be treated during their absence from France as the others previously sent to Morea.

Some good boatmen will also embark on the *Luxor*, and they will be very useful in the course of your operations.

You are also authorized to take on locally the number of men that you judge necessary to employ as laborers; our current information leads us to believe that you will find always as much as you desire and at a very low price.

The French consul general in Alexandria will put at your disposal the funds that will be necessary; it would thus be good for you to request them from him and to keep an exact account of all your expenses, to justify on your return.

The intervention of this same consul general will procure for you very great facilities from the Egyptian authorities in obtaining the modes of operation that you will desire, and even if necessary help against all kinds of opposition. You have in addition at Alexandria a natural supporter in the reputation of Mr. de Cerisy, officer of the French Naval Engineers, officer in charge of directing the naval constructions of the Viceroy of Egypt; I have no doubt that you will find him most eager to give you good advice, and to help you with the means of execution that he has at his disposal.

I add here copies of several documents, the knowledge of which may be useful to you. These are, first, a memoir of Mr. Besson, former officer of the French Navy, now in the service of Muhammad Ali, on the means of transporting the obelisks of Luxor. Second, a letter of Mr. de Cerisy on the same subject. Third, an extract of which I have already spoken above of Mr. Taylor, and finally the letter of Mr. Champollion, which we mentioned at the beginning of this letter.

I believe, sir, that I have omitted nothing that might help to ensure the success of the mission with which you have been charged; your talents and enthusiasm are for me guaranteed by your prior services, and I am confident that you will succeed completely in the operations you will execute.

Kindly send me reports of the progress of your operations as often as possible.

I attach here also a letter for Mr. Mimaut, France's consul general in Egypt, and another for Mr. de Cerisy; they are intended to announce the mission with which you are charged, and to ask them to help you with all their ability in the matter of execution. Please deliver them on your arrival in Alexandria.

Signed, Count Sebastiani

Chapter 2

Departure from Toulon and arrival at Alexandria — Interview with Muhammad Ali — Departure from Alexandria — Rosetta — Aspects of the Nile and its banks — Arrival at Cairo — Krali Effendi — Departure from Bulaq for Upper Egypt — The Pyramids, Memphis — Sherif Bey — Dendara — The flotilla arrives in sight of Thebes

The *Luxor,* loaded with all the usual apparatus for military maneuvers, got under way from Toulon on April 15, 1831, under a favorable northwest breeze and reached the coast of Sicily in five days.

After departure from Sicily we sailed with quartering or following winds. Under those two speeds the vessel behaved very well, reaching speeds of up to eight knots. Things were different when, during the day of the 21st, the winds veered to the east, and forced us to tack. The *Luxor* was not made for sailing into the wind, making a lot of leeway—as might have been expected as the inevitable consequence of its required design. We spent the night of the 24th–25th with a high sea running that put a lot of stress on the mast. The following day, however, the wind having risen and returned to the northwest, we continued our route without further complications or problems, and on May 3 the *Luxor* docked at Alexandria port, opposite the palace of the viceroy. Several minutes later Mr. Cardin, chancellor of France, was on board. He informed us that Mr. Mimaut, consul general, was in Cairo along with the viceroy, who lived in that city during the bad season. The presence and help of Mr. Mimaut would be indispensable for me to put into operation the plan that I formed during the crossing. I planned to leave immediately for Upper Egypt, to precede the *Luxor* to Thebes, to explore before the inundation the land we would be operating on, to prepare an anchorage to receive the vessel, and at the same time to commence preparatory work. All this was forcibly suspended till the arrival of the consul in Alexandria, which took place twenty days later. This unfortunate beginning of the campaign seemed to me a bad omen.

However, I devoted the delay to sightseeing in this city, whose monuments serve as signals to the sailors and guide the Arabs in the desert. All has been said about Alexandria, about the scattered ruins, her ground disturbed and in turmoil in all directions; it is a subject that is exhausted for scholarship. All that remains is for the traveler to see, admire, and be silent.

Among the ruins of the ancient city, the needles called Cleopatra's naturally attracted my attention. Often I would go to visit them to gain inspiration from their location, to study, to meditate the issues that I had taken in hand to solve, and to prepare by thinking through the procedures I proposed to solve the difficulties that would be offered by the movement of such a considerable mass; but I was empowered by the novelty of this operation, the confidence, the energy, and the perseverance necessary to surmount them.

On June 8 the pasha received us in his palace situated on the Pharos.* Told in advance by the consul general of my small stature, he pretended not to see me, and asked with a laugh, "Where is the engineer?" On my presentation by Mr. Mimaut he replied, "He was hiding behind you. Tell him to sit beside me so I can see him." This situation also gave the viceroy the chance to make several pleasantries that showed in him a lively imagination, a mind quick to grasp reports.

His eye, always quick and scrutinizing, lets nothing escape it. His finesse, his perspicacity, the depth of his ideas, his accurate replies give him an advantage over all who approach him, and which it would be difficult to avoid.†

Muhammad Ali, passionate about great enterprises and in general about anything out of the ordinary, showed us several times the lively interest that he had in the success of our bold project.

"It interests me," he told us, "as if it were carried out in my name and for my glory. The most formal orders have already been given so that nothing that needs to be contributed to the accomplishment of this gigantic work will be refused to you."

The viceroy then addressed to me, and also to the captain, various judicious questions relative to the fleet built by Mr. de Cerisy, the naval engineer; in a word, the generous and noble conduct of Muhammad Ali, his

* In the time of the pharaohs, the Pharos was a little island that emerged from the water two miles from the coast. This reef was joined to the town founded by Alexander by a jetty that divides a vast harbor into two parts. At one end of the jetty, on the Pharos, rose the famous lighthouse built by the architect Sostrates under Ptolemy Philadelphus. It is on this dike, now considerably widened, that the modern city has been built.

† It is a pity that this prince, endowed with such rare qualities, sacrificed the interest of his people to ambitious projects that he could not complete. Humankind will ask him to explain all the good that he could have done and all the bad that he did do. It will also demand an account of his behavior toward the sad remains of the Egyptian race, whom it would have been well to civilize instead of destroying them through violence, and whose story would make anyone feel pity.

eagerness to assure me of all desired facilities and the help of local officials, inspired in me the greatest confidence in the mission that was entrusted to me.

From this moment, nothing kept me any longer in Alexandria. The supplies had been transboarded from the *Luxor* onto local boats called *djerms*,* all was ready for my departure, which circumstances outside my control had prevented till now.

On June 11, with the rising sun, the transport boats set sail for Rosetta; I embarked with all the workers on the largest of the *djerms* that composed our little fleet. Toward four in the evening, the yellow color of the water and the frowning face of the skipper announced to us the approach to one of the mouths of the Nile. This opening is obstructed by the mounds of sand that are displaced daily. The flow of the current, combined with the violence of the sea, creates through these obstacles one or several openings, frequently winding and of uneven depth; the path that is considered navigable is named *boghas.*† It is always lined with rocks, against which the waves often break with force according to how the wind agitates them. God help any boat that finds itself enveloped on these breakers. It runs the risk of being thrown onto the coast and thus perishing. Soon we observed two *djerms* upside down, partly covered with silt, indicating previous shipwrecks. Our boats, however, crossed this dangerous passage without accident; we thus entered this superb river after having gone around an island which did not exist forty years ago. It was, on the contrary, the deepest point, where the cannoneers of the French army anchored in 1799. Three hours later we were in Rosetta, in the middle of the freshest and greenest of all countries.

This town, formerly so rich, so flourishing, that used to be one of the principal centers of commerce of the Orient, has languished in a sad decadence since the opening of the Mahmudiya Canal.** This canal, whose entrance is located three leagues upstream, joins the Nile to the old port of Alexandria. All the products of Egypt are distributed by this path, which requires less expense and offers more safety than transport by sea.

* One sees on the Nile three kinds of boats: the *agabas*, the *djerms*, and the *maschs*. The first ones are heavy flat-bottomed boats that draft little water. They are used to transport merchandise during the season of low waters. The djerms, whose construction is more favorable to speed, go up the river with a lot of speed thanks to their huge lateen sails. Finally, the masch could be called the sailing barge of the Nile. These are 400- to 500-ton boats that only navigate at the time of the inundation.

Besides these transport boats, the Nile is dotted with two other kinds of boats, the *dahabiyas*, which carry light merchandise and passengers, and the *canjiehs*, narrow dinghies whose lines are very fine and which resemble the yacht in their elegance and lightness.

† In the Orient, this term applies to every narrow and difficult marine passage.

** This is the former Cleopatra Canal. Muhammad Ali used 150,000 Arabs to build it in several months. Twenty thousand died from the heavy work. It was called Mahmudiya in honor of the reigning sultan.

And so the population of Rosetta has considerably diminished, to judge by the great number of houses that are not occupied.

In this place, materials still had to be transported on *agabas*, flat-bottomed boats especially made for navigation on the Nile. Meanwhile, the *Luxor* went through the *boghas* and came to anchor in Rosetta, where it had to stay until the swelling of the river permitted it to continue on this route until Thebes. A few sailors were divided among the *agabas*. Mr. Jaurès, seaman first class, and Mr. Pons, physician's aid, embarked with me on a *canjieh*, a light, narrow dinghy that goes at the greatest speed under sail or with oars. A small wooden shack rose in the back of the boat. It is divided into two rooms; the first serves as a living and dining room, the second is a bedroom. All these preparations being completed, we said goodbye to our fellow travelers who remained on board the *Luxor*, and on June 19 the flotilla, leaving Rosetta, started to go up the Nile.

At this point our attention was mainly stimulated by the objects surrounding us. All was new for us, the appearance of the places, the trees, the plants, the clothing and the language of the Arabs who were running in big crowds on the banks, and for whom we were an object of curiosity. Our eyes would rest agreeably on the delta, on this land of new creation whose numerous and various products are for the most part unknown in Europe. Not far from Rosetta, on the left bank, is the famous mosque of Abu Mandur, a place of pilgrimage where women come looking for fecundity, which they attribute to the waters of a fountain within this religious monument. Beyond, the two banks, lined with reeds, open to the astonished eye of the traveler forests of palms, mimosas, rich crops; but next to those fertile fields are vast deserts without any appearance of vegetation. Here and there one still sees vestiges of a beautiful country fallen into the most horrible poverty, of great villages in ruins, of fertile lands lost to cultivation, bad huts built of sun-baked bricks or, more often, mud. The upper part of these houses, if one can give them that name, is reserved for pigeon coops from which myriads of pigeons escape. The murdering gun never scares these peaceful inhabitants. The Arabs do not eat them; they only take care of them for their droppings, which make an excellent fertilizer.

Under these mud huts covered by palm leaves vegetates an unfortunate population which has for law, for social institution, nothing but the caprice and the stick of a few Turks often more brutal than animals. You see everywhere men covered only with rags, exhausted by hunger, aged by bad treatment and deprivation. Children of both sexes entirely naked beg for assistance from the traveler. At the least signal from a Turk, every Arab is subjected to the kurbash.* According to this arrangement there is perfect equality in the domains of

* Whip of hippopotamus hide; see part 2, chapter 2 (Voyage in Nubia). This is how the Turks inflict punishment with the whip.

the viceroy: everything there is submitted to a terrible, and I repeat, stupid despotism.

Our navigation up to Cairo presented several difficulties that I was far from expecting. The *rais*es or bosses of the *abaga*s used to exploit one of these conditions of their trade, just as in Paris the coachman exploits the traveler who pays by the hour. Since it is to their advantage to make the trip as long as possible, they ran their boats aground on sandbanks on purpose. This maneuver, which delayed progress considerably, caused me acute worries. Already the greenish color of the waters indicated the rising of the river. It was necessary to arrive at Thebes as soon as possible to prepare, before the inundation, the dry dock on which depended in part the success of the operation. Hurried by this consideration, fearing most that some part of the material would be lost, of which the slightest piece was indispensable to me, we took advantage of the speed of the *canjieh* to proceed forward, backward, right and left, on all the points where an *agaba* had run aground. We forced the crew to haul in the deep water, sometimes helping them the best we could, sometimes being helped by the inhabitants of the local villages.

One of the ships, the one carrying the most important materials, the *filin-blanc*,* hit a sandbank and lost its helm. The shock caused by the impact created a considerable leak. The rope was going to get wet, damage which could have the gravest consequences. There wasn't on board either pumps or instruments that could be used to drain the water, and it was only by the use of their leather hats and pieces of clay pottery that the sailors and workers were able to get the water out of the hull. Once this operation was over, the carpenter and caulker put the ship in a state suitable to navigate to Cairo. We arrived in this town on June 27, 1831.

Immediately I went to the Citadel, where the governor lived, as much to pay him a visit as to ask for the intervention of his authority. The first thing he did, according to the intermediary, was to give the order to cane the *rais*, according to the law of the land. I opposed this cruel punishment. However, to prevent malice from the latter, I asked the governor to have four *cawas*† embarked on the flotilla who would be charged to watch the *rais* during this journey. His Excellency responded graciously that he could not refuse me anything and at the same time gave me a letter addressed to Krali Effendi, director of navigation, by which he authorized the official to put at my disposal a new *agaba*.

Upon leaving the palace we went down to the well, said to be Joseph's, which did not offer anything remarkable. It is dug in the rock and at a rather great depth. The water was raised through several wheels to which are attached earthen cups. I was already regretting having given several minutes to my curiosity when our

* Cord used to make the ropes of the tackle.

† A kind of janissary.

interpreter, Yusuf, a former French Mamluk, brought to my attention with his usual nonchalance that it was already time for every good Muslim to retire to his harem and that the director of navigation could only be seen the following day in the afternoon. Whether we liked it or not, it was necessary to put up with this forced delay.

The audience was set for four o'clock. Krali Effendi, seated on a kind of stand on the bank of the river and protected against the strength of the sun by several tents, was giving orders to the Nile sailors. As soon as he saw me, he motioned for me to sit next to him. This was a man of high stature, of Herculean force, common face, and whose hard and imperious glance made all the Arabs tremble.

After the customary salutations, I explained to him the reason that brought me to him. Instead of answering me, he told us that he had been put in charge by the pasha to transport "the stones of the king of France to Alexandria." (This is what he called the obelisks of Luxor.) "Upon arriving there," he told me, "I was scared by the size and mass of those monuments. It was not at first glance, but only after having made precise calculations, that I drew back from such an exercise whose execution seemed impossible."

To prove his point, he showed me a ruler divided into equal parts, bearing ordinal numbers indicating the volume and weight of the horizontal section of the monolith corresponding to a height proportional to each division (scales on the ruler) with a total weight of about five hundred thousand, "an immense load." He added, "You must try to move it and bring it down from its base without accident."

Surprised by the knowledge and correctness of the calculations of Krali Effendi, I asked him several questions about the obelisks with respect to their positions to the river and ruins and their condition. He replied that the obelisk to the right when entering the temple was streaked by a crack, which starting from the base went up to about one-third of the height. "That is not possible," I exclaimed, turning toward the guide. "You misunderstood. Only the pyramidion is chipped." My exclamation, my surprise, made the director smile. He got up to show me on one of the poles of the stand a crack almost similar to the fissure in the obelisk. He indicated, however, that he could not have checked whether this crack went through the granite because the opposite side is hidden by homes where they confined the women, a haven always respected by the Turk.* Taking the small notebook that I had in my hand, he started to sketch, on one of its sheets, a camel mounted by an Arab of the desert. If these observations of the director had surprised me, I was even more surprised

* There was in this expression a certain oriental exaggeration, because even if this were indeed true for "the stones of the king of France," it is no less certain that in other circumstances and for other reasons, entire villages are destroyed in a few hours by the soldiers of the pasha, without any considerations for the women.

at seeing what he had just made under my eyes.* The camel was well done, the features of the Arab perfectly characterized.

Krali Effendi had learned the elements of drawing and geometry under the direction of Mr. Coste, a French architect, and he had rightly retained a high idea of the talents of his master. After this meeting, which lasted more than two hours, he gave the orders to the portmaster to put at my disposal one of the best Nile boats.

To come back to the crack that Krali Effendi had mentioned to me, it is remarkable that it was not mentioned elsewhere, neither in the great works on Egypt nor in the more recent report by Champollion the Younger. On the contrary, one finds there that the shaft of the right obelisk is in a state of perfect preservation. I forced myself to believe that it really existed, and almost enjoying the doubt, I tried to convince myself that Krali Effendi had confused the right with the left, in which, in fact, the lower part is indeed cracked, or that he had come up with this pretext to excuse the failure of his mission to the pasha.

I thought that I would encounter no more obstacles, but it wasn't so. The day after my meeting with the director of navigation, the portmaster claimed that, because all the *agabas* were loaded, the one that was to be mine would only be available in three days. Informed by experience of the greediness of the pasha's servants, I slid into his hand a few *kiries* (small gold coins). The touch of this metal suddenly awakened his diligence and solicitude, and soon put an end to this problem. One hour later, I was in possession of the coveted *agaba* and, thanks to another use of the same persuasion, I received the necessary men to make the necessary transfer.

On July 30, in the evening, the flotilla left Bulaq for Upper Egypt. We were finally going to reach the desired goal. It was then that, reassured by the presence of those janissaries on our boat, I decided to take a quick trip to the pyramids of Giza. These monuments, whose aspect is so grandiose, so imposing when one sees them from the banks of the river, lose some of their grandeur and some of their proportion as soon as one climbs the plateau on which they rest. The reason for this is that their tapering forms and angularity hide them from the eye, and after having crossed the cultivated land, one does not find any object for comparison. Whatever it may be, one's judgment gets lost in the presence of these regular masses that rise in the middle of a vast desert. But if one lists the number (203) of tiers of large stones which make up each of these huge pyramidal staircases, if one compares the height of an average step to the size of an ordinary man, if one thinks that, seen from a short distance, the top seems to end in a point, though in reality it has a surface equivalent to forty square meters, then

* As is well known, the Islamic religion disapproves in principle of all painted or drawn representations of objects or people. This disapproval, it is true, has little force in Egypt today, since there is a school of drawing at Abu Za'bel. But nevertheless, it still is rare to meet a Turk who knows how to handle a brush at all deftly.

all the attention focuses on the unusual dimensions that only imagination can fathom. Their true positive values, reduced to numbers, will give for the side of the base 233 meters, for total height 146 meters, and for the volume, 2,662,628 cubic meters.

We went inside the largest pyramid through an opening made about a third of the way up the west side, forming an inclined passageway. Guided by torchlight through these ancient chambers, we went through, with an admiration that caused silence and stupefaction among us, long galleries and two chambers, the greater of which holds a beautiful sarcophagus. Never had anything so grand struck my eyes. It is a gigantic piece of work that seems to go beyond human efforts. The quantity of material employed in the construction of the pyramids, the difficulty of positioning these blocks, both exterior and interior, the durability of the solidity of the immovable masses, which are as durable as the mountains that border Egypt, the hundred thousand workers who worked there for more than a century, place them among the greatest human enterprises.* What patience, what constant work necessary to coordinate so many immense details! The pyramids, whatever one says, are living proof to present generations of an antique civilization without parallel, just as they will be for centuries to come, a civilization able to find within itself sufficient resources to raise those vast tombs and men capable of conceiving and realizing such a project.

On leaving the pyramids, we went toward the plain where Memphis is believed to have flourished. According to tradition, this emplacement was crossed by the river, which the first pharaoh, Menes, diverted, as if it were possible for the first king of a country to execute such an operation. We will have the opportunity to return to this topic in the course of this book. There only remains of this ancient city a great chain of mounds covered by palms and heaps of broken stones piled up all over. However, one can see by inspecting the place that this city was perfectly described by Herodotus and Strabo. Memphis was limited to the north by the pyramids of Giza and to the south by a line of smaller pyramids more than two leagues in length.

Who could walk this soil, freshly watered by French blood and former witness to our triumph, without remembering with admiration the energy and courage of our soldiers? It was there, said our Arab guide, that the French army, moving en masse, pulverized the best cavalry of the universe (Mamluks) in spite of the *nar kitir* (strong fire) that was burning their feet. And further on, does not this cloud of reddish dust contain the wandering souls of the brave of Heliopolis?†

* Champollion, following Manetho, attributes the building of the pyramids to the pharaohs of the Fifth Dynasty. If one agrees with the chronology of this historian, these monuments go back more than four thousand years before Christ.

† Translators' note: During Napoleon's Egyptian campaign many French soldiers lost their lives at the Battle of Heliopolis.

June 30 was devoted to this excursion. A twelve-hour walk under the beating sun had exhausted us. Our poor dog jumped on the sand screaming in pain as if he had put his paws on hot coals. He would have undoubtedly died from his suffering if one of us had not placed him on his donkey. At six in the evening we arrived at Saqqara, a city situated on the banks of the Nile, where the *canjieh* was waiting for us.

In a hurry to return to the flotilla, which had to carry us about fifteen leagues, we continued to go upriver, where fertile waters bathe a wide valley, bounded right and left by a chain of undulating white mountains, without any kind of soil. During the day the heat is excessive, unbearable for a European, but in the evening the navigation of the Nile becomes very pleasant. The northwest breeze refreshes the atmosphere; stars barely visible in Europe give their radiance under a sky always pure and peaceful. Gathered in a circle around that big mast, the sailors sing Arab songs, and the boat, its antennae* overlapping, went so fast because of its speed through the waters of the river that it caused the water to bubble up, and on it there were long shadows of the forests of palm trees and sycomores. All this created in the soul a calm and delicious feeling of admiration.

In front of us passed several hamlets or villages such as Atfih, Beni Suef, Abu Girga, and sometimes villages covered in green, but sad, poor, and whose silence is only disturbed by wild screams and bitter complaints. Here, as in Lower Egypt, despotism holds all of the population under a brass net whose every link is the center of action of a tyrannical subaltern. As Montaigne said, "To have several masters is the same as being extremely unhappy the same number of times." Civilization, no doubt, will penetrate one day; it will infallibly be born from the contact of the Occident and the Orient. But let us not give this name to the virulent and brutal organization that still weighs on this unfortunate country.

Until this point, the Nile does not offer in general any of the inconveniences of our great rivers. One does not find in its bed either rocks or stones; even a pebble is a rarity. Going upstream is always possible. The northwest or northeast wind present during all of the flood provides the means to defeat its rapid current through the simple help of the sails. Only in bends does the breeze which was favorable in the original direction become unfavorable. Then the sailors jump on land and haul the boat with a rope, regulating their steps with a rhythmic song. During this maneuver the progress of the boat was so slow that we could have, without wasting much time, made an excursion toward the mountain, where many things were attracting our attention. There were excavations, openings in the form of doors dug into the side of the Arabian chain that borders the Nile, the pillars, the columns, supporting porticoes with still vibrant colors. But I knew by experience that the hours go quickly for the traveler stopped by these gigantic

* The lateen yard where the sail is attached.

ruins thrown with such profusion and majesty on the Egyptian soil. I had thus restricted myself to landing only at Asyut, in order to give to Sherif Bey, governor of Upper Egypt, the letters I was carrying. This forced break happened earlier than expected, but it took place in Mallawi.

Several boats were anchored in front of this town, situated about two miles from the riverbank. We confused them from afar with our transport boats. At a closer distance the interpreter recognized, from the flags flying at the tops of the masts, the boats of Sherif Bey. His Excellency was touring his governorate and had to stay in Mallawi for several days. I arrived there after the sun set. I had myself announced to the governor for very early in the next day.

At six o'clock I went with the interpreter toward Mallawi. We went through a part of this town that is larger and better built than the previous ones. From north to south it is crossed by a large bazaar, which seems to indicate a trading and industrious people. But this meeting place, this hearth of oriental life, was almost empty. Most of the shops were closed, the houses abandoned, everything pointed to a significant diminution of the population and economy.

Ibrahim, the interpreter of His Excellency, of whom we will have the opportunity to talk several times, invited me into the apartments of the governor. They had hidden the quite miserable condition of the room he was in with the kinds of ornaments with which they decorate boats. The parquet was covered with mats and rugs. A chair, crudely decorated, was placed in front of Sherif Bey. The interpreter invited me to rest there immediately because such is the custom of the country. It is considered polite to sit down to breathe for several minutes even before paying respects.

His Excellency, seated in the oriental manner, his arms softly extended on elegant velvet cushions, was surrounded by several officers, his doctor, two subalterns, and several Arab servants. After a great number of *salams* (salutations) and compliments, which lasted no less than five minutes, they served the pipe and coffee. The moment had come to give to Sherif Bey the orders of the pasha. He read them very attentively, turned toward his officers and the doctor. He pointed out to them with an air of satisfaction and contentment the last sentence of the letter. It was a compliment of Muhammad Ali, honoring his faithful servant with the name "his son."

The face of Sherif Bey indicated a quick mind, finesse, and a disposition to happiness. The sound of his voice was remarkable by its softness. Far from having the arrogance and brutality of his compatriot, Krali Effendi, he distinguished himself by a gracious welcome, by those polite and natural manners which were a sign of good company. Named, although very young, to the government of Upper Egypt through the sole favor of the pasha, he maintained himself there by his intelligence, his zeal and limitless devotion. Careful to second the views of his sovereign, Sherif Bey never missed any opportunity to learn what could contribute to the accomplishment of the vast projects of Muhammad Ali. He was also convinced that the French, naturally very communicative, must not have any more secrets 1,100 leagues from their country, so he submitted me to a long questioning on European diplomacy and particularly on the plans of Russia, a world power that he seemed wholeheartedly to detest. Learning that I had taken part in the Algerian campaign, he wanted to know the slightest details of this memorable campaign. After I had answered all his questions, and in a way that seemed to satisfy him, we came to speak of the obelisk.

Sherif Bey told me several times of his desire to be in Luxor on the day the monolith would be taken down from its pedestal. He then gave me letters for the *nazirs*[*] of Qena, Qus, and Esna, informing these officials of the object of my mission and telling them to help me with all their power. So the meeting ended as it started, with mutual cordiality and satisfaction.

Having left Mallawi at noon, we reached the dangerous passage of Abu Feda at 9:00 p.m. This mount, which is part of the Arabian chain, rises steeply one hundred meters up on the right bank. Its sides offer strangely shaped outcroppings of which it would be difficult to make an exact picture. The Nile, forced to deviate from its course by this natural rampart, rolls its muddy waters along this gigantic shelf with the greatest speed. Reflected by this obstacle whose crown is topped by sharp needles, the winds react in all directions and storm with violence.

The flotilla had passed Abu Feda before the day ended and was going in good order toward the next stop. The captain of the *canjieh*, less courageous and perhaps more superstitious than his friends, and further scared by the darkness produced by the shadow of the mountain, refused to navigate these difficult narrows in the night. In spite of his observations about the dangers menacing us, we forced him to go forward. Taking his knife, he planted it in a crevice of the main mast, threw a handful of salt into the fire, put his fate in the hands of the Prophet, and reassured by this strange ceremony, he adjusted his sails.[†]

* A name given to one of the secondary agents of the Egyptian government.

† This ceremony reminds us of a solemn sacrifice done by the Chinese sailors before crossing the Yellow River. The captain of the boat, surrounded by his crew, tears off the head of a rooster and douses with the blood of the victim the deck, the mast, the anchors, and the doors of the rooms, to which he attaches a few feathers of the same animal. Then, invoking the river divinity, he pours oil, tea, an alcoholic liquor, and some salt on the surface of the water.

It seems that in all times and places, salt has always been considered a necessary element to sacrifices. The law of Leviticus commands the Israelites to salt all offerings. According to Ovid, in all the offerings of the early Italians the law was *Puri, mica sale* (Purity through salt). Horace, speaking of offerings to angry household gods, mentions the *saliente mica.*

The *canijeh*, fighting painfully with the aid of all its sails against the current of four to five knots, had barely passed a third of Abu Feda when a strong gust of wind fell upon the boat like a bolt of lightning and cracked its mainmast. This was the precursor of a true hurricane. We went instantly from a complete calm to a violent breeze that went around the compass in five minutes. Thrown in those tumultuous movements, our frail boat advanced, sometimes forward, sometimes backward, to the right, to the left, rushing under its two sails, more often under the storm jib, and sometimes with useless sails. The more we advanced, the more the storm redoubled in intensity. Carried by the violence of the wind up to the foot of the mountain, we were made aware of our critical position only by the crashing of the waves against the reef, when suddenly we saw on the water a light that disappeared intermittently from sight. The scene, now lit, was only more frightening. As the flickering light of this fire reflected on the walls of a vast excavation, one could distinguish part of the mount, which seemed to threaten to fall upon the navigator. Then the steep rocks, pointing their shining heads in the bed of the river and against which the winds and current struck, were pushing us with the greatest impetuosity. The danger was imminent. The *canjieh*, enveloped in eddies, rolled like a straw along those rocks and almost broke several times. It was only after maneuvering for two hours with great care and skill that the Arab sailors managed to get through this most dangerous spot. Then the wind calmed a bit. The gusts became rarer and soon we reached the end of Abu Feda. Three hours later we reached Manfalut, where the flotilla awaited us. Our stay in this town was not long. A brisk northwest breeze made us leave early in the morning, to the desperation of our *rais*, who, since the passage through Abu Feda, had developed a horror of navigation.

Our progress, always favored by a regular wind, was so quick that after five days we were in sight of the superb forests of palms and doums that preceded the ruins of Dendara. This latter tree, of which I had been told in Lower Egypt, could be of great utility in carpentry work. I had asked Sherif Bey for authorization to cut as many as I would need. His Excellency gave me an order to this effect for the *nazir* of Qena, to whom I went at 7 p.m.

When the tumult of the first compliments and salutations subsided, I gave him the governor's letter. In order to avoid any difficulty, I told him right away that the price of the doums would be paid immediately to the owner according to the estimate made by the local authorities. "With money," he said to me, "we will send you a whole forest to Luxor." After having three cups of coffee, according to the customs of the land, and smoking as many pipes, I took leave of the good old man. My surprise was great when upon exiting his house I was seized by some *sayises*,* who put me on a horse whose enormous size contrasted comically with my short

* *Sayises*, domestics serving as runners and often as personal valets.

stature. My legs were horizontal. As my predicament seemed very inconvenient, the head of the runners tried to reassure me by boasting of the docility of the horse, which his master, old and crippled, did not fear mounting while smoking his pipe. However, as a measure of precaution, he placed on either side two *sayis*es holding the tips of my boots, while two others held the bridle and tail and four domestics holding torches preceded us. They took me in a procession to the *canjieh.* I later learned I was given this great honor, for which one pays dearly, because one must pay generously all those who take part. It would be bad for someone to come empty-handed; he would be poorly received. (And this is not the case only in Egypt.)

We reached Qena. This town of about fifteen thousand souls is the meeting point of all the caravans of pilgrims that go to or return from Mecca. Although very ancient, it offers no vestiges of antiquity. But Qena is renowned for its excellent pottery. A porous clay is found there, used in making *alcaraza*s (clay vases used to cool water), which are used a lot in Egypt. It is from here that every year, rafts of jars, pots, etc., are sent to the main cities, where these potteries are sold at a low price. In front of Qena one finds the ruins of Dendara, the first well-preserved temple that one meets on the route we had just taken.

In spite of our extreme impatience, we had to pay a just tribute of admiration to this colossal work of Egyptian genius. During the night, the flotilla crossed the river. As the sun rose, officers, sailors, and workers seized by an electric curiosity went running toward Dendara, as if the ruins might escape them.

I will not attempt to describe the sublime spectacle offered to our eyes. What language could produce sufficient terms to paint the effects of this marvelous monument? Such a task is above my strengths. The temple of Dendara in itself says more than thousands of books and drawings. Everything here is grand, indescribable, prodigious.

I shuddered on noticing on one of the partition walls traces of the barbaric ignorance of the Turks. To the eyes of those barbarians, antiquities are but heaps of stones, a quarry ready to supply materials partly fashioned for their crude constructions. Already the hammer of demolition resounds on the ancient paving stones. The building would have been mutilated, destroyed, annihilated forever but for the intervention of the European consuls, who prohibited this barbaric act. The temple was preserved, respected by the Turks, but civilization was not as kind. It is from the ceiling of Dendara that they removed the zodiac that today is found in the Louvre.

The diversion of this strange and instructive stop impressed on all our faculties new energy for new labors. Each one of us felt amply compensated for the fatigues of the trip by the sight of these impressive ruins.

The following morning, at the break of day, the flotilla under sail was slowly advancing toward its destination. A light breeze, and numerous detours, delayed our progress. Finally, the Qamula elbow was crossed by means of a towline toward four o'clock in the evening. Then from all sides appeared the gigantic vestiges of the ancient city. Lit by the last rays of the sun, those ancient masses offer an admirable spectacle, witness to the creative power of a people that no other has equaled in their grandeur of vision.

To the right, the ruins of Gurna, the Memnonion. Behind, the Valley of the Tombs. A little further at the foot of the Libyan chain, the temple of Madinat Habu. To the left, those of Karnak, its immense pylons, its giant obelisk thirty meters high. A little further south, the colonnade of the palace of Luxor, its huge pylon preceded by two granite obelisks. At the sight of these two monuments everything was erased, disappeared from my eyes. For the first time the sight of the ruins of Thebes awoke in the soul of the traveler an idea other than historical monuments, of greatness and decadence. Thebes! It is no longer the city of a hundred gates with these objects, palaces, the sphinxes, colossi. This is but the spot that encloses one single object, the obelisk to the right upon entering the Ramesseum.*

The lead boat was going toward the bank of Luxor and came to dock in front of the obelisks. The others imitated its maneuver. At the same moment, sailors and workers rushed onto the land, crossed the shore running, and disappeared among the ruins. The *canjieh*, forming the rear guard to keep an eye on the stragglers, arrived and placed itself in the center of the flotilla.

* Translators' note: Lebas uses the term "Ramesseum" because the obelisk and pylons were built by Ramesses the Great. Today, "Ramesseum" refers to the mortuary temple of Ramesses, which is a different monument on the other side of the Nile.

Chapter 3

Arrival in Thebes — First examination of the obelisk; the stone is cracked — Demolition of the homes encircling the monolith on the right — General aspect of Luxor — A description of all the monuments — Reconnoitering the localities — Some historical details

The first duty of an engineer being to give an account of everything relative to his work, my object here will be to reproduce the facts relating to it, the series of impressions under which I found myself placed, in the order they presented themselves. The reader must not expect a complete and methodical description of the places and monuments, a historical treatment of their origins. . . . Later perhaps I will get into some details regarding this topic, but such was not the preoccupation that seized me at my arrival in Thebes. For me, I repeat, everything was in that mass of granite weighing around 230,000 kg, which needed to be uprooted from the ground and loaded on board a ship destined to transport it to France. From my point of view, there was something grand in this operation that exalted my soul, but it was mostly the question of mechanics, the difficulty of the task, the solution of the problem, and above all, the responsibility weighing upon me by a mission given to me in the name of France.

It was therefore not without emotions that we set foot on the soil that was to be the stage for our works. For more than an hour people had been waiting for us with impatience.

The sight of our boats bearing the French colors at the tops of our masts, the hope of being rewarded a few paras* from the Europeans, had attracted on the shore some of the inhabitants of Karnak and Luxor. There, as in all of Egypt, we were assailed by a crowd of poor people asking for baksheesh (alms). After those cries of misery that matched the rags barely covering their nakedness, the Arabs, encouraged by the gift of a few

* Local money.

coins, came to gather information from our interpreter about the purpose of our voyage.

The idea of taking down an obelisk and loading it on a ship for France seemed to them so bizarre, so outlandish, that in spite of the assertions of the interpreter, they persisted in believing that we had other intentions. According to them, we had come into the Said to explore the places, to probe into the minds of the people, in order to come back later and take over the country. "The obelisk" they said, "is obviously a false pretext. You know as well as we do that these *birbe* (antiquities) were built here with a special cement that hardens in the sun, so it is impossible to move them without demolishing them, cutting them to pieces. Does there exist, besides, a ship big enough, strong enough, to carry this enormous load? And what man would dare to undertake such an operation?" "Here is the man" answered Yusuf, pointing toward me. The answer of the interpreter made all those present burst out laughing. "Allah, Allah!" they yelled. "Who? That one? But my stick is taller than he is. He is not capable of moving even the smallest stone of the temple."

During the conversation we were going toward the obelisks. Having crossed several mounds of debris and an avenue of gigantic columns, we entered a winding passage flanked by modern constructions. Our presence among the inhabitants made a crowd of children of all ages come out of the huts, screaming "Baksheesh" (alms; it is, I believe, the first word that children learn to speak in this country.) This road ends at two pyramidal masses united at the lower part by a brick wall pierced by a little door. One goes in through this opening onto a square where the obelisks and colossi that decorate the façade of the palace of Luxor are located.*

The sailors and the workers had preceded us in this visit. The Italian, Mazacqui, a stonecutter, was gently tapping the east side of the obelisk on the right, and was paying close attention to the noise resonating from the mass. As soon as he saw me, he cried out in his language, half Italian, half French: "Moussou [Monsieur], the stone, it is cracked, but I do not believe it is broken. The sound is good. We will be able to remove it, providing that it falls slowly, very slowly." The problem indicated by this worker was obvious. One had to agree with the evidence and recognize that, as rude as he was, Krali Effendi had been right in his observation.

I returned to the boat, tired, deep in thought, embarrassed at having examined nothing, not even the road we followed to return on board. I was walking like a drunken man, incessantly repeating, "There was a crack that not a single book on Egypt had mentioned."

* Translators' note: Lebas repeatedly calls Luxor Temple a "palace." This error goes back to Napoleon's Egyptian campaign. When the scholars Napoleon brought with him saw battle scenes on the walls of the temples, they could not imagine such scenes being placed on their churches or cathedrals and concluded that the buildings were palaces. When they published their monumental *Description de l'Égypte*, all the temples were called "palaces," and Lebas continued this error.

A thousand contradictory thoughts assailed me at once, full of turmoil and agitation. I was at the same time in Toulon, Paris, and Thebes. The sole thought that I might be accused of shattering the obelisk while taking it down from its base, or while taking it on board, was absorbing all my faculties.

The following day, after a few hours of rest, I found myself in a totally different frame of mind. I was almost ashamed of my excessive worrying. "Once the helmet is on your head," said Juvenal, "it is no longer time to go back." I put myself to work. I toured the exterior and the interior of the temple to get acquainted with its various parts and to make myself a topographical map containing the obelisks, the ruins, the flow of the river, and the shape of the land. The result of this examination was that it would be totally impossible for me to apply the plans I had conceived in the course of the voyage. It is a truth about plans conceived in an office that they always fail because of a few essential data—an unforeseen peculiarity in the soil, in the distances to be covered, or in the surrounding objects, and suddenly the scaffold that one had laboriously conceived and raised comes crashing down.

I thus had to draw up a new plan to adjust the procedure to the new circumstances, to carefully combine all the parts of an apparatus that, because of the nature of its function, might be instantaneously destroyed by the rupture of one of its parts. We would have to decide definitively the givens of the problem, deduce from it the equations, trace exact measurements, and submit the whole to the rigorous test of calculations to guarantee results.

As early as the day following our arrival in Luxor, the interpreter had assembled a few fellahs* who, ever since, had been working to clear the foot of the obelisk, as the shaft was partially covered in sand. I was impatient to uncover it entirely in order to be able to see if the crack went down to the base and went through the granite. This work and the subsequent operations required the demolition first of more than thirty houses, or to speak more exactly, thirty huts. As miserable as they were, however, the sight of gold, usually so powerful on the Arabs, could not convince them to sell. I didn't know then that they were trying to gain some time in order to devise some means to take away in the night, and without the Turks knowing, the goods that they usually steal in the fields at harvest.

Instructed about these difficulties by a messenger on a camel carrying a report dealing with the circumstances, the governor of the Said sent to Luxor his interpreter, Ibrahim, and the Cawas-Bashi† with a formal order to execute immediately everything that I would request. They, upon their arrival in Thebes, wanted to end the discussion *à la Turque*—that is, so to say, using the whip. I strongly opposed it. It was

* Peasants.

† A kind of officer of the pasha.

agreed that a commission presided over by the *nazir* and composed of the *qadi* (the head of the village), the Cawas-Bashi, and the two interpreters, would decide the payments to give to the owners. This commission met on July 25, 1831.

Answering their requests, the Arabs gave a list of the houses to be demolished, followed by an estimate of an amount more than ten times the real value. According to Ibrahim, all present seemed strongly impressed. The youngest complained that they would be expelled from their paternal roof without giving them either time or means to build another house. The old people explained in a supplicating tone of voice that it was wrong to think they had money, and that they had no other means of survival than their little pigeon coops. The oldest one, whose head was adorned with a green turban, which is a mark that he was a descendant of the Prophet, maintained a grave silence, but contracted his lips while his hand was clenching convulsively at the end of his foot (he was seated cross-legged), revealing the agitation he was feeling. The chief of the village and the qadi (judge), his incandescent face sweating, insisted upon the exactitude of the measuring stick with which the measurements had been made in the presence of the *muhandis* (engineer). One poor blind woman dressed in rags humbly asked for two *thalaris*[*] for total payment. Her accent, her calm, were proof of her sincerity. I immediately gave her three times what she asked, with the promise to replace her hut with a brick house. To the contrary of what one would expect, this act of humanity produced a very bad effect. The Arabs persisted more than ever in their exaggerated claims and they based their evaluations on the house of this poor blind woman.

After a long discussion about the price of a *pic*[†] of brick masonry, and on the length of the stick that had been used to measure, Ibrahim put the final end to the discussion. Indeed, standing in the middle of the enclosure, waving his arms and legs, we could see his long and thin face, characterized by the enormous protuberance of his nose, take successively a grave and menacing air, addressing with force each owner and exposing the government's displeasure against the village. There was an outburst of guttural sounds mixed with various inflections, of piercing screams impossible to describe. In the end, this petulant Armenian produced more words and more gestures in a quarter of an hour than a Turk ordinarily makes in an entire year. Taking finally a calmer, more moderate tone, Ibrahim placed himself alternatively next to the oldest Arab and the qadi, spoke to them in their ears, made a thousand grimaces, got angry again, and then listened to them with goodwill and ended up by promising a baksheesh, a bribe, an argument which never fails to produce a

* 10 francs, 40 centimes.

† A measure of length in the country.

magical effect on the Arabs. At the same time, all at once, the families concerned in the sale of the houses came together in front of the commission. The women were showing only their eyes and foreheads, upon which could easily be read the symptoms of pain and anger. They let out a few reproaches and their sobs kept them from continuing. This incident caused the *nazir* and the Cawas-Bashi, who up to now had smoked their pipe without saying a single word, to cry at the same time "Rou-barra" ("Get out of here").

Finally, after many altercations and clamors, and right when everyone was screaming at the top of his lungs, I learned to my great astonishment that the debates were closed and the business was done. "We are in agreement," said His Excellency's interpreter, "and the *nazir* invites you to dinner."

At these words, the Arabs calmed down like the waves of the sea under the terrible *quos ego*.* Each one of them upon leaving came to kiss my hand, a proof of respect and affection used in all of Egypt.

The interpreter brought me to a room decorated with a couch that dominated the area. In the middle was a circular tray supported by an elegant pedestal table. There were no knives, nor glasses, nor plates, nor forks. Before sitting at the table, a servant placed on the knees of each guest a napkin decorated with a gold braid. A second one came holding in his hands a ewer and a metal basin with a bar of perfumed soap. After washing hands we sat down, or rather, we gathered around the pedestal table, a position quite uncomfortable for a European, clothed as I was in my formal uniform.

The dishes, in general well prepared, were served with order and method. One dish never arrived before the previous one had disappeared.

Conforming to etiquette, I had to taste before anyone else. I had barely touched a dish when the hands of the guests dipped avidly into the one dish, placed in the center of the table. People took meat with their fingers, tore it with their teeth, and dunked the bread in the sauce. Each dish remained only a few minutes on the table and was immediately replaced by another. I counted thirty-five, among which were two whole sheep. These animals were cut up by the *brin-bashi* with much dexterity. He then offered me the most delicate piece, which he was squeezing in his hand like a sponge. A second tray covered with sweets, creams, fruits, rice cakes, etc., followed the first.

Throughout the whole meal we only drank Nile water served in small jars from Qena. Wines and liquors are not permitted in these fancy meals.

Dinner finished, the ewer and the basin were again circulated. This time, considering the way we had just eaten, this precaution of cleanliness was a necessity. Then a servant offered me a very beautiful pipe ending in a piece of amber decorated with several precious

---

* Translators' note: These words were spoken by Neptune to the disobedient winds in the *Aeneid* (I:*135*).

stones. The *nazir*, wanting to pay me the greatest of honors, did not smoke. The other pipes, ten in number, rested on a pipe rack fixed against the wall for everyone to see. Several minutes later I offered it to my host. It was a duty of politeness that I had to fulfill as seriously as possible. In his turn, he passed it successively to the main guests. Meanwhile, coffee was served. It was only after several rounds of the pipe and coffee that I could decently take leave of my host.

The following day I paid to the owners the agreed sum. In order to avoid all future difficulty, we made two copies of the discussion regarding the selling of the houses, and a copy was given to the agents of Sherif Bey. They were waiting for this piece of paper to go back to Asyut. According to the customs of the Orient we had to shower them with presents and thanks. We shall see how well they were deserved.

I found myself several hours later with the interpreter of the expedition, the former French Mamluk, about whom I have already spoken. I was already praising the intelligence and zeal with which his colleague had just removed the obstacles hindering the execution of our work when Yusuf answered with a smile, "Ibrahim is but a juggler. With a single word he could have solved the problem. He only had to say that the government wants it, and the whips would . . . . But in trying to prove that his help was indispensable to you, he wanted to secure for himself a gift directly proportional to the service rendered. He always acts that way with foreigners and I can give you the proof. He told me upon your arrival: "Well, Yusuf, for how much do we sell this gentleman?"

So, all this incredible scene of the selling of the houses, threats on one side, pleading on the other, all was a ruse and a deceit. It was all about getting as much ransom as possible from us. This would explain thus the silence that our interpreter, Yusuf, had kept regarding the matter, when we found out that he had taken advantage of the circumstance to get rid of, at our cost, a gun decorated with inlays that the governor's interpreter had already agreed to accept as a proof of my gratitude.

Upon writing this account of my voyage, I am allowing myself to give a few details which in the end will seem perhaps superfluous to the reader, or even beneath the gravity of the matter. But, I must say, putting aside the interest they can have, it seemed to me that they did belong in this account in order to make known my true predicament in the country, and the customs of the men among whom a special mission had taken me.

During the three days that our problems with the Arabs lasted, I had busied myself with leveling the ground, taking exact measurements in order to start the urgent work.

These preliminary details do not offer any difficulties in a temperate climate, but in Luxor, in the middle of the desert whose sands burned the feet, under a July sun which, in four hours, made the Réaumur thermometer climb to 53 degrees,* one needed more than just goodwill to spend entire days surveying the land. Fortunately, I found unexpected help in the zeal and intelligence of Mr. Jaurèss, a first-class mind. This young man, endowed with a quick intellect and a great preciseness of execution, enabled me to finish this work in good time. Since during my stay in Thebes I had no other help than his to manage and watch the four hundred Arabs employed for the cleanup, a great part of the credit for our results belongs to him.

I also owe more than one useful idea to the kindness of Mr. Bernard, chief engineer of bridges and roads, detached from the service of the navy to assist us. It is to his enlightened advice, fruits of a long experience, that I owe my success and my lack of discouragement. I always remembered that, after a long discussion on the system of demolition and containment walls, Mr. Bernard told me these words that the facts have justified: "The success depends mostly on the details of the execution that are dependent upon the locations. It is only on site that you will be able to make a definitive plan. The task is difficult, but I am convinced that you will complete it perfectly."

I will now proceed to the examination of the ruins of Luxor, and in particular of the ground on which we had to operate, and whose principle parts I depicted on Plate I.

Luxor, the most beautiful village in the area, and whose population of around eight hundred souls was composed of Arabs, Egyptians, and a few Copts, is situated on a mound 700 meters long and 350 meters wide. This artificial tell is three meters above the level of the plain, and contains huge ruins, monumental columns, obelisks, and colossi that eclipse the objects around them. What one notices, above all, are pyramidal projections preceded by two pink-granite monoliths that make up the entrance to the palace of Luxor. The houses of the village, built of red or sun-baked bricks, surround the north side of this edifice. Some are enclosed within, and others, in particular the house of the *Brin-Bashi*, were built on the roof of the palace. Most of these dwellings have on top of them a pigeon coop crowned with tree branches where clouds of doves come to roost.

The southwest corner of the palace points to the bosom of the river and is protected from the shock of the waves by a brick quay of substantial thickness. This quay, which in modern days has been repaired and lengthened by an extension of stones taken from ancient constructions, slopes down toward the Nile, away [from the palace] in blocks seven meters long. Each inundation causes more and more serious damage. The soil is removed and the bend becomes more and more pronounced. It is probable that the Nile will end up turning this corner by digging away to the east of the construction.

* Translators' note: This is the equivalent of 66 degrees Celsius, 151 degrees Fahrenheit.

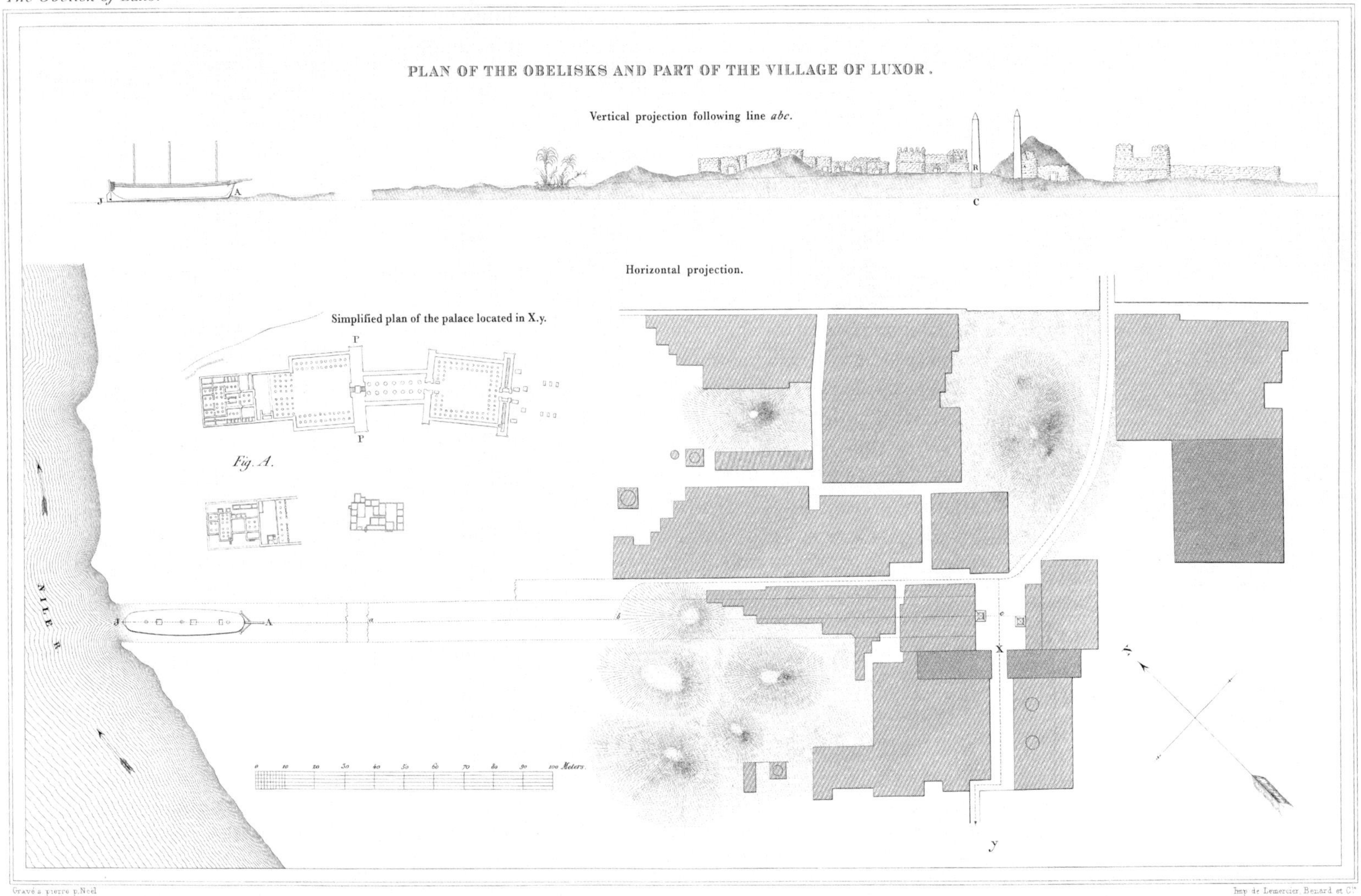

Plate I: Plan of the obelisks and part of the village of Luxor

Such is the general aspect of the antiquities of Luxor, of which we will give a more detailed survey. This examination is indispensable to understanding the work that preceded the taking down and loading of the obelisk.

We will take as a starting point a staff, J, raised on the right bank of the Nile in alignment with the two obelisks (Plate I). From this viewpoint, one sees only one obelisk. Following the direction of this alignment west and east, one walks on a beach of empty sand, basically horizontal, that terminates in the mound of the village. There one is suddenly stopped by a small steep hill of debris about ten meters high,* crowned with the shacks that we just discussed. Since this hill is situated between the bank and the obelisk, to get near these monuments one is forced to go north around its base on a small winding path which exits in a street of Luxor parallel to the departure path. At the end of this street, turning to the left, one enters a rectangular space enclosed on three of its sides—at the back by the wall of the palace, on its sides by a double row of houses. Two houses abut a side of each of the obelisks. These monoliths protrude from this square, whose ground is level with the beginning of the hieroglyphs. A brick wall pierced by a little opening replaces the seventeen-meter-high door that was originally between the massive walls. This opening leads to the middle of a big square filled with modern constructions, which conceal the remains of ancient constructions. One sees only a few sandstone blocks with hieroglyphs in front of the mosque, which was built at the place where the vestibule used to stand. This part of the building is buried 5.9 meters below the debris of fragments of bricks, pottery, and a large quantity of sand brought in by the desert winds. A door made to the right in a dirt wall gives access to a winding alley, lined with modern houses that lean on the columns of the vestibule. This alley leads to a gallery composed of fourteen gigantic columns with capitals in the shape of an inverted bell. It was then easy for us to recognize the road we had followed on our first visit to the obelisks. We had entered the palace on one of its sides.

The wall P marked on the plan is entirely destroyed. One can barely see a few remains to the east, at the level of the debris. The only remains of the next wall are an architrave and the back colonnade. It is there that the hospital for the sick of the expedition was set up. Next comes a square vestibule encumbered by sand and supported by four columns. This apartment, covered by large flagstones, was used for housing workers. On the right and at the end there used to be a door giving access to a lateral room located on the side of the water that is today demolished, as well as the surrounding wall. Next comes a sanctuary connecting through a door in a transverse corridor, full of debris. We chose

* The Arab builds as little as he can. Should a house fall down, he will build another one on the same place. He doesn't bother to take the trouble to clear the debris. This is what has caused, in large cities, the small hills and even mountains of debris that one does not notice at first.

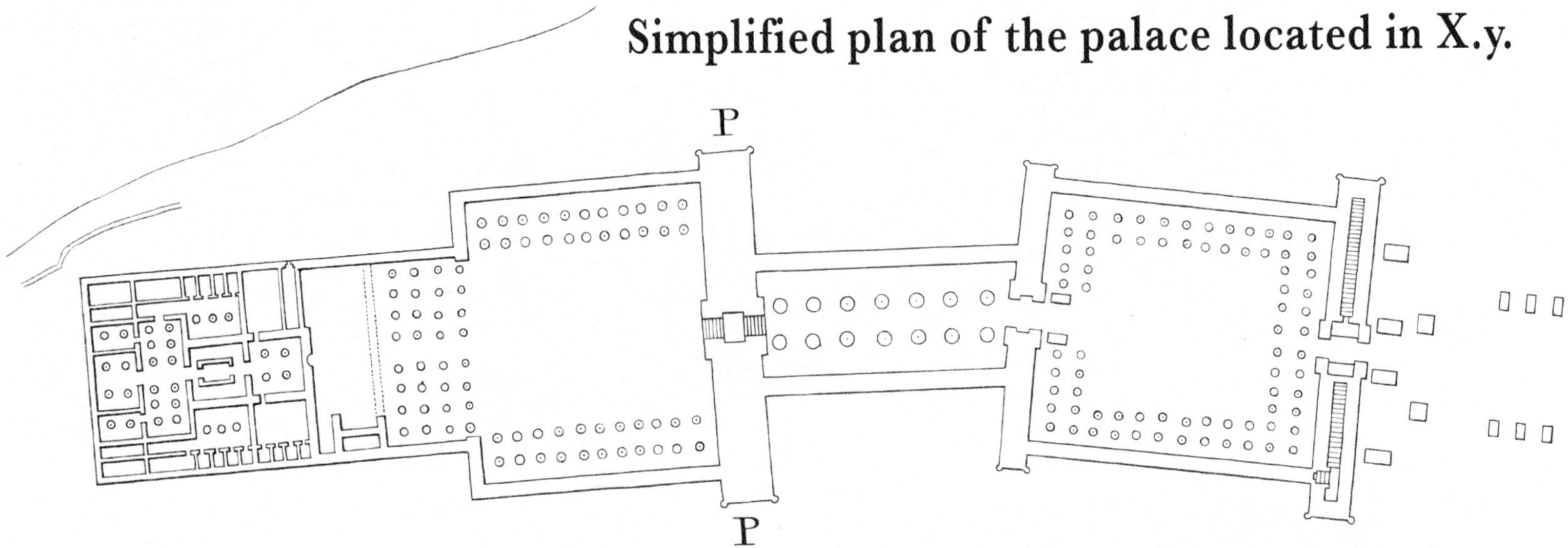

Detail from Plate I: Plan of Luxor Temple

this gallery, which is surrounded on all sides by apartments, to house the crew. Two of the adjoining rooms served as storage rooms for the food and supplies, and we built an oven in the third one located in the southwest corner.

Leaving through this exit, we climbed on a path along the south side of the building. This path leads to a horizontal road, parallel to the east side and dug through the embankment. It is almost level with the beginnings of the capitals of the columns. A stone stairway placed perpendicularly serves as a means of connection between this road and the top of the palace, where the habitation of the *Brin-Bashi* was located. It is composed of a dining room and three other rooms. On the orders of Sherif Bey, this officer was quick to give me the use of these rooms. Since they were not sufficient to lodge the administrative staff of the ship, the construction of several rooms, a reception room, etc., had to be undertaken. All these little houses, which we will mention again, rise on the part of the roof between the south enclosure wall and the north end of the vestibule occupied by the workers. Let us return now to the obelisk square.

Following a direction opposite to the first one, one enters the main street of Luxor. It goes through the entire length of the market and village. This was probably the old route that connected the two quarters of Thebes on the east bank of the river. The remains of sphinxes that one finds on this path up to Karnak tells us that it was lined with them. To the left upon exiting the square was a steep mound of debris, and opposite in front, on the east side of the road, an area enclosed by earth walls. This place, close to our work, was perfectly suited to serve as a depot for the materials brought from Toulon. It belonged to a holy man who gave it to me at a very inexpensive price.

Since the historical circumstances that are tied to the buildings of Luxor were discovered on these very grounds by Champollion the Younger, we do not believe we can do better than to quote from a letter written on the subject by this famous archaeologist on March 25, 1829. It is dated from Biban al-Muluk [the Valley of the Kings], a hamlet in the territory of Thebes.

> Our work at Luxor ended (almost completely) before we came to settle here in Biban al-Muluk. I am able to give all the necessary details about the time of construction of all the parts making up this great edifice.
>
> The founder of the Palace of Luxor, or rather of the Palaces of Luxor, was the pharaoh Amenophis Memnon (Amenophis III) of the Eighteenth Dynasty.* It is this prince who built this series of buildings stretching south to north, from the Nile up to the fourteen great columns, 45 feet high, and whose masses also come from this reign. On all the architraves of the other columns adorning the courtyard and the interior rooms, columns numbering 105, for the most part intact, one reads in large hieroglyphs, of low relief and of excellent work, dedications made in the name of the king, Amenophis. I give here the translation of one of them to give an idea of all the others, which only differs in a few titles.
>
> > "Life! Horus powerful and moderate, ruling through justice, the organizer of his country, he who holds the world steady, because, in his great strength, he battled the barbarians. The king (lord of justice) beloved of the sun, son of the sun (Amenophis), moderator of the pure

* Amenophis III ruled around the year 1680 BC. It is he whose statue, according to the ancients, had the marvelous gift of emitting harmonious sounds at sunrise.

> region [Egypt], had these constructions built, dedicated to his father, Amun, the lord god of the three zones of the universe in Oph [On] of the south. He had them built in good and hard stones in order to erect a durable edifice. This is what the son of the sun, Amenophis, beloved of Amun-Ra, has done."

These inscriptions remove all doubt about the exact era of the construction and decoration of this part of Luxor. The fourteen great columns mark the end of the work of the reign of Amenophis; however, under him, columns 2 and 7 of both rows going from south to north, were also decorated. Their bas-reliefs belong to the reign of King Horus, son of Amenophis, and those of the last four belong to the following ruler.

The whole section of buildings in Luxor north of the fourteen columns is from another era and forms a separate monument, although it is connected by the great colonnade to the Amenophion, or Palace of Amenophis. It is to Ramesses the Great* that we owe these constructions, and his intention was, not to embellish the palace of Amenophis, his ancestor, but to build a separate edifice. This is made clear by the following dedication inscription, sculpted in large hieroglyphs below the cornice of the pylon and repeated on the architraves of all the colonnades that the modern shacks have not yet buried.

> "Living! The Aroeris [Horus] son of Amun, master of the Upper and Lower regions, doubly kind, Horus, full of strength, friend of the world, the king (Sun, guardian of truth, approved by Phre), the preferred son of the king of the gods, who is seated on the throne of his father, ruler of the land, had all these constructions built in honor of his father, Amun-Ra, the king of the gods. He built this Ramesseum in the city of Amun, in Oph of the south. This is what the son of the sun (the beloved son of Amun-Ramesses)† has done, the eternal giver of life."

---

* This prince is none other than the famous legislator and conqueror Sesostris, who ruled 1,550 years before the Christian era.

† These words in parentheses indicate here, as in the inscriptions already cited, the contents of the cartouches: the prenomens and proper names of the kings.

It is thus a separate monument, distinct from the Amenophion, and this explains quite well why these two great buildings are not on the same alignment, a shocking defect noted by all travelers, who wrongly supposed that all these edifices came from the same era and formed a single entity, which they do not.

As for the subjects sculpted in bas-reliefs on the pylon of the temple, they are also of great historical interest. The vast surface of each of these two masses is covered with sculptures of a very fine style. The subjects are all military and composed of several hundred human figures. The right one: the king Ramesses the Great seated on his throne, in the middle of his camp, receives the military chiefs and foreign envoys, details of the camp, luggage, tents, wagons, etc., etc. Outside, the Egyptian army is engaged in battle, war chariots in the front, in the back, and on the sides. In the center the infantry in military formation. Left wall: Bloody battle, defeat of the enemy, their pursuit, crossing of a river, the taking of a town, the prisoners are brought.

Here is the general topic of these two tableaux, of about fifty feet each. We have very exact drawings of them as well as of the few inscriptions intermingled with the military scenes. The large texts relative to this campaign of Sesostris are beneath the bas-reliefs. Unfortunately, one would have to demolish part of the village of Luxor to have copies of them. I thus had to be satisfied with learning from the upper, but visible, lines that this war had taken place in the fifth year of the reign of the conqueror, and that the battle had taken place on the fifth of the month Epiphi. These dates proved to me that it was indeed the same war whose events are sculpted on the right side of the great monument of Abu Simbel, which also bears the date of Year 5. The battle represented in this temple is also in the month of Epiphi, but Day 9 and not 5. It is obviously two battles of the same campaign. The people that the Egyptians had to fight are Asiatics, who can be recognized by their clothing as Bactrians, Medes, and Babylonians. The country of the latter is explicitly named (Naharaina-Kah, the land of Nahraina, Mesopotamia) in the inscriptions of Abu Simbel, as well as the lands of Schot, Robschi, Shabatoun, Marou, Bachoua, which are surely to be found in the early geography of Western Asia.

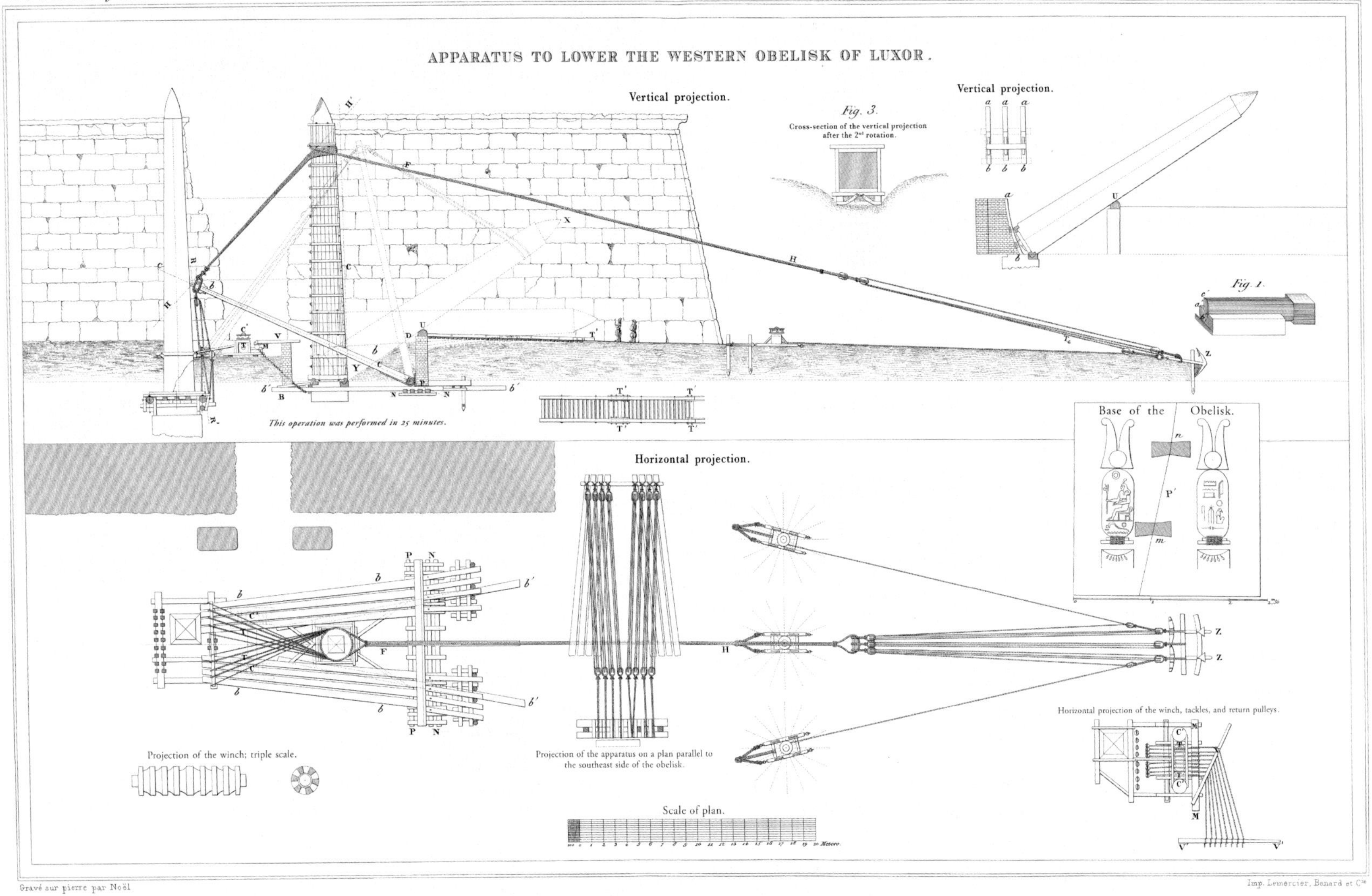

Plate II: Apparatus to lower the western obelisk of Luxor

Chapter 4

Preparatory work for lowering the obelisk — Description of equipment — Outbreak of cholera — Continuation of the work — Uncovering the base of the obelisk — History of the obelisks of Luxor; text of the inscription on the Paris one

If we have succeeded in giving, in the previous chapter, an adequate idea of the localities, and the position of the obelisks with respect to the pylon of the palace, to the village, and to the river, the reader will understand without difficulty that the most simple means, the quickest to arrive at the target, consisted of laying down the monolith on its western surface, the closest to the bank, and to haul it aboard. As a consequence of this first arrangement it was necessary to run the boat aground in the direction of the two obelisks, and align its three masts with the monuments and with the surveying rod, J, from where we started measuring; then we had to pull down all the houses situated on the debris mound, excavate across these ruins a reasonably sized road capable of maneuvering the apparatus, and establish another road communicating with the bank and the street of Luxor which runs by the square of the obelisks, to bring the materials there. This was the full picture of the work of the installation of the machinery for lowering and restraining, as well as the construction of the accessory buildings, that I could foresee.

For a long time all my illusions concerning the crack had vanished. The fifteen Arabs employed to excavate the foot of the monolith were put to work to uncover the eastern face. The blemish descended to the lower edge and seemed to penetrate the granite. Nevertheless, this excavation provided me with an essential measure: the base of the monument, which rests on a pedestal, was buried 3.8 meters in the ground of the square. It was no longer a question of where to run the *Luxor* aground; the only thing left to know was the approximate high level of the river. I obtained this with the help of traces left by the Nile on various parts of the bank. These indications agreed with the accounts provided by the inhabitants.

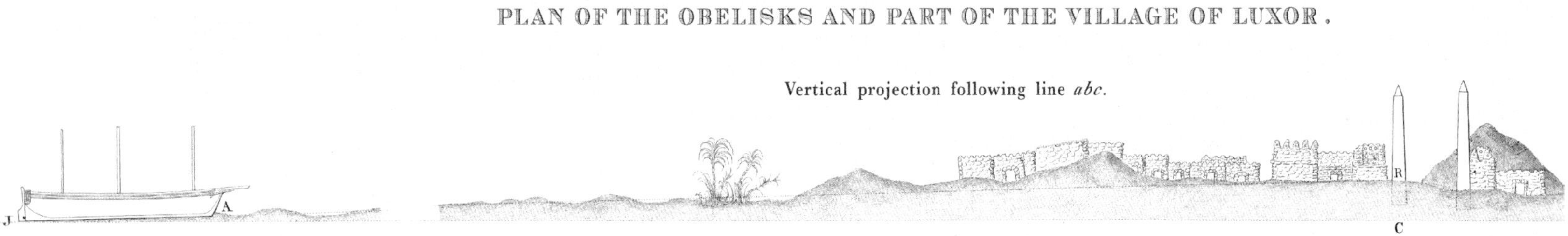

Detail from Plate I: The *Luxor* run aground and aligning with the two obelisks

The contours of the topography indicated that it would be necessary to take advantage of the actual inundation in order to position the boat four hundred meters from the obelisks. Then, after the retreat of the waters from the land, the boat would have to remain there until it was raised by the new flood and afloat once again, with a cargo of 250 tons. The part of the riverbank (AJ, Plate I) where the boat would be run aground was found to be within several centimeters (in elevation) of the top edge of the socle. This circumstance permitted us to excavate a horizontal road passing by these two points, onto which the monolith could be lowered and subsequently hauled on board the ship. But a work of this nature, along the full length of the road, would have required a mass of debris in excess of 45,000 cubic meters. Quite apart from these difficulties presented by the transport and excavation of this immense volume of sand, and the time necessary to do it, it was feared we would encounter at this depth some debris, old constructions, ancient or modern, unforeseen obstacles which would be impossible to predict. After thorough examination, I came up with another idea; I imagined lowering the obelisk completely by two successive rotations.

After having cleared and prepared the ground as we have indicated on Plate II, we proposed to incline the monolith toward the bank by turning it on one of its edges of the lower square, up to where it would hit the stub of pivot U, where the distance to the socle was calculated in a manner to correspond, after this (tilting) operation, with the center of gravity of the mass [being over the pivot]. Having arrived there, the obelisk would be in equilibrium on the roller so it would be easy to start a tipping movement about this cylinder; this gyrating movement would simultaneously lift its bottom and lower its top on the road higher than the top of the pedestal. It would thus result in a considerable saving in excavating, equivalent in volume to the prism JCR in Plate I, and the advantage of an inclined plane that would facilitate the transport of the monument up to the ship.

Firmly agreed on this foremost point, we still needed to devise the means of execution. This consisted of two systems of apparatuses. The first, which we call the *lowering apparatus*, was intended to incline the monolith toward the bank, pulling its summit down. The *restraining apparatus* served the same function, but in reverse. It would function at the moment when the vertical of the center of gravity of the mass passed the axis of rotation. Its action was intended to regularize, to moderate at will the speed initiated and communicated by the accelerative force of the weight, which, without this precaution, would undoubtedly lead to the breaking of the monolith. This is what happened to the obelisks of Rome, when the barbarians pulled them off their pedestals. The installation of these two apparatuses demanded preparations that we will explain.

After assuring myself that I could immediately engage several hundred workers who would either prepare the ground, or take down the houses, or finally unload and carry the materials into the enclosure of which we have spoken, I announced in the various sections of Thebes, and in the local villages, that the workers employed as diggers would each receive, without intervention by the authorities, a daily salary. This news was very well received by the inhabitants. The hope of receiving a fair salary instead of the lashing of the whip brought to Luxor a great number of workers.

The Nile rose rapidly; because of this, the preparation of the bed for running the ship aground had to be done right away. Nothing was neglected to prepare the ground according to the slope as designed.

Despite all the desirable enthusiasm, the water reached the workers, whom we were obliged to protect by movable dikes that were constantly shifted from one point to the other. We were able to judge by this initial work the difficulties that we would have to surmount to excavate the roads for hauling through the mounds of debris. Although they were used to making excavations or uncovering the bottoms of monuments, the Arabs were unskilled at cutting the ground following a determined inclination, an operation that required the application of measuring sticks, leveling rods, and cords, which they were completely ignorant of how to use. It was also necessary to follow them step by step, to indicate the points where they deviated in digging to make trenches and embankments, etc. However, five days were sufficient to excavate a channel fifty meters long by twelve meters wide with a depth of 1.2 meters. During this interval, the boats were unloaded, and the materials were brought to the enclosed worksite near the obelisk, which from then on became our magazine. The workers took possession of the vestibule, which they excavated and arranged conveniently to receive the materials. New workers arrived in the area and joined the first ones, and eventually the whole length of the shore between the obelisks and the grounding bed was occupied by four hundred workers. All was movement on this line, where a cloud of burning dust rose that remained suspended in the air, hiding from our eyes a part of the village.

The workers were composed of men, women, and young children of both sexes. The first were charged with pick-axing, the others with removing the debris. Their manner of working seemed interesting enough to be reported.

On the right side of the road were the boys, on the left the women and girls, divided into three groups. Each group serviced a suitable number of pickers, waiting for their baskets to be filled so they could take them away. Then, when each one had helped his neighbor to place the load on his head, the signal to depart was given by heads of the right-hand columns. They sang a song, very obscene, which their comrades, clapping their hands in rhythm, repeated by heart. The young girls imitated the same motions, and responded to the boys with similar chants and accompanied by the same gestures. With these chants were intermittently mingled Arabic cries from those in charge of directing them: "*Yalla, volet! Yalla, benti!*" ("Let's go, boys; let's go, girls.") The parade took place in front of the overseers, who threatened each worker with a blow from a palm branch, but they avoided it with much skill by quickening or slowing their step. These ways that are repugnant to our European customs should not be surprising in a country where it is customary to mention by name all the unmentionable parts of the body and all their functions.

Well habituated to living under a sky that we would say is always on fire, and that reminds one of the strength and exact expression of Moses ("furnace of iron"), the Arabs need only a small pair of shorts, the women a simple cotton robe; children of both sexes were entirely naked. At certain hours of the workday, the heat was so intense that the young ones, whose feet were less hardened, could not take the heat of the sand; they carried their baskets at a run, dumped them out quickly, and put them under their feet for several minutes.

Soaked with sweat, made thirsty by the salty dust they breathed, these unfortunates came constantly to quench their extreme thirst in the muddy waters of the Nile. Water was kept in jars at various points of the worksite and near the workers. At the sound of a drum, which announced the end of the work, men, women, and children all hurried to the river and dove in for half an hour. After their bath, they reassembled on the shore. The overseer was in the center of a circle whose circumference was occupied by forty or fifty workers sitting cross-legged. I would go from group to group and distribute to each worker the salary that had been allotted. The poor blind woman was always present at this distribution. "Give" she said, "to the unfortunate; she cannot help in the work, but she will pray to God for you. You will be successful."

On August 13, the chief of the boats that go upriver to Philae came to us announcing the arrival of an immense vessel. Stricken by the grandeur of the proportions of its masts and sails, some compared it to a floating mosque, others to a *feddan* of land.* But what was even more

* *Feddan*, an agricultural measure of the country. [Translators' note: equal to about 0.42 hectares or 1.03 acres.]

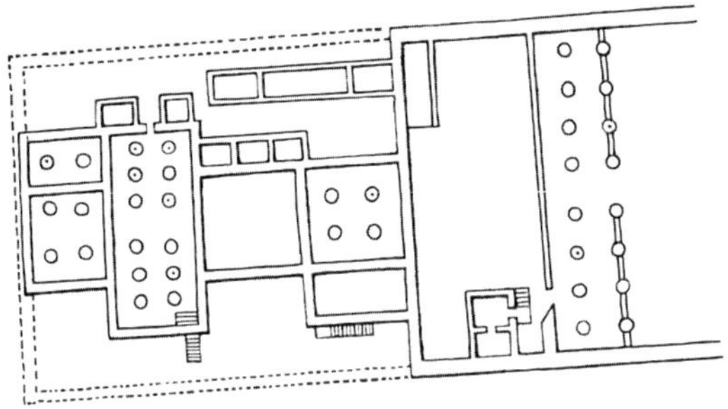

Detail from Plate I, figure A: The housing built at Luxor Temple for the workers

astonishing was to see this mass maneuvering with the aid of only a small whistle. It wasn't necessary to shout, sing, or say a single word. This was the *Luxor*, which docked on August 14, 1831 in line with the obelisks. Later it was placed in the dock that had been prepared to receive it.

From this moment, it was necessary to devote ourselves without delay to clearing the two rooms of the temple, one where we intended to house the equipment, the other to serve as a hospital for the sick. A great number of our workers were immediately applied to this clearance. At the same time, the masons were working to restore the roof that time and vandalism had destroyed. The trunks of date palm trees cut into three parts acted as beams, replacing the architraves and the ancient paving stones. We filled holes with branches from the same tree. On top, we applied mud from the river mixed with chopped straw to form a floor.

On top of this platform we raised in a few days the officers' chambers, a divan, a reception room, a larger terrace, etc. Below and in front of the temple, other rooms for the chiefs, the storeroom, the bakery, etc., and all that we have related in Plate I, figure A were created. The parts of the temple that no longer exist are indicated by dotted lines. The whole thing, well repaired and whitened with lime, gave the village a new look.

These little houses were built according to the custom of the country, with bricks dried in the sun and from river mud, mixed, as we have already said, with chopped straw. They formed, in fact, a mass of mud that in our western climates the rain, or even the humidity, would melt in a few hours. But, under a sky without rain or dew, where the air is always perfectly dry, these hovels can last a large number of years. Besides, this manner of building is very old. We read in Exodus that the pharaoh, irritated with Moses, cut off from the Hebrews the chopped straw which formerly had been provided, without diminishing their quota of bricks. This problem was one of the causes of the Israelites' flight in the desert.

While these works were executed with all desirable promptness, I stopped others no less important and relating to the apparatus for lowering and restraining. The carpenters, guided by the foreman, Elias, whose enthusiasm and intelligence I must praise, did not remain idle.

On August 1, three ladders were placed end to end, rising vertically along the south face of the obelisk. At a given signal, the national flag floated on the top of the monolith. Sailors and workers saluted with cries of "Now it belongs to us. It belongs to France; Paris will soon own this monument." I was quick to answer their acclaim with two lines of La Fontaine:

> . . . You should never
> sell the bear's skin without first killing it.

After taking possession in this way, we got ready to record all the dimensions of the obelisk, and to set up the work site as well as workshops for the forge and finishing area. From that day on, all the workers were occupied. The Arabs cut the two mounds of debris and demolished the houses, and others built new houses on the roof of the edifice. The carpenters framed the obelisk in a scaffold and climbed to the top to take measurements; the sailors prepared the pulleys; the master carpenter ran back and forth, placed material, counted the pieces of wood, adjusting the framework, and recommended the sparing use of materials that seemed insufficient for the work to be done.

It would assuredly be forcing the comparison to say that at the sight of this population, bustling around the monument, one would think oneself back in the time when the ancient Egyptians raised these wonders of architecture. But we can at least confirm that perhaps for many centuries the silent ruins of ancient Thebes had not witnessed such movement, such a frenzy of work. Here, it is true, it was not about building; we were going to remove one of the most imposing remains from the last vestiges of an ancient civilization to enrich a new civilization. Before giving an account of this operation, it is necessary to set out the mathematical aspects of this problem, and to enter into a few details regarding the system and the projects to be executed.

For the moment we will consider the western obelisk, the one on the right as you enter the temple, as simply an immense block of pink granite. The operations to which it must be subjected have as their goal to take it down and load it onto a beached ship.

Its shape is that of a quadrilateral truncated pyramid with a parallel base. The little base topped by another pyramid, called a pyramidion because of its short height, tops the monolith. The pyramidion is noticeably damaged, with a crack that originates from the base and streaks the two faces east and west up to a third of the length. The side closest to the river faces the west; the one that faces the pylon of the temple faces the south; the sides symmetrical to the previous ones face to the east or north.

Dimensions of the large base:	North edge	2.44 meters
	South edge	2.42
	West edge	2.42
	East edge	2.42
Dimensions of the small base:	North edge	1.5 meters
	South edge	1.5
	West edge	1.58
	East edge	1.58

Height of the truncated pyramid:	20.9 meters
Height of the broken pyramidion:	1.94
Total height:	22.84

From these figures, one deduces that the volume of the pyramid is 83.46 cubic meters, and that of the pyramidion is 1.54 cubic meters, which gives a total of 85 cubic meters, of which the weight is 229,500 kg, using the estimated number of 2.7 for the density of Syene granite.

Upon close examination of the western obelisk, by putting rulers on the sides and edges, it becomes clear that the sides are not planes. The west and east sides have a double curve, they show a crosswise convexity that the plumb measured at 3 centimeters for the first and 3.5 centimeters for the second. This fact has been confirmed and explained by the members of the Institut d'Égypte, but since those scientists did not have the use of a measuring compass as we did on all the parts of the monuments, the second curvature must obviously have escaped them. This curvature along the length of the same sides is all the more remarkable in that it is turned in the opposite direction of the axis of the monolith. In other words, the west side is convex along the length while the east is concave on the same axis. As a result, the four longitudinal edges are curved lines whose concavity is turned toward the east. The line of curvature of the western edges is 0.02, those of the eastern edges is 0.045.

It would be rather difficult to explain this singularity. It cannot be attributed to chance or defective work, because the eastern obelisk also presents the same peculiarity. The four longitudinal edges also present their concavities toward the east.

I return to my subject.

Whatever means was used to take down the monolith from its pedestal, it was necessary to clad it with an envelope to preserve the sharpness of its edges, the polish of its surfaces, and the carvings that are engraved into it, and finally to protect it from an array of other possible accidents during transport, unloading, and placement in Paris. In the system that we had adopted, it was also necessary in addition to protect from any fracture the edge of the base that would serve as the axis of rotation. We attained this double goal by the following means.

All the sides of the monolith were covered with timbers, or in nautical terms, planks, 0.16 m thick, held at intervals by horizontal casings. These casings were formed of two oak crossbeams, notched 0.05 m in the planks, and two bolts that pressed them against the granite. The crossbeams and the bolts were placed alternately on the four sides to tighten the planks of wood equally. By means of these connections, the various parts of the system were locked to each other, fixed immovably. But the whole of the framework could slide up and down because of its pyramidal shape. It was urgent to prevent this movement and to protect, at the same time, the lower edge. It was with this in mind that the crossbeams of the last encasement were hollowed out, following interior angle ABC. This framework, placed under the base of the obelisk and tied to the other crossbeams, formed a stopping point and would prevent skidding from bottom to top. Since the outside of the joist in which the axis of rotation lay had been rounded off, the only thing left to do would be to put it into a container dug to the same curvature, to complete the hinge around which the gyration must take place (Plate II, figure 1).

A close look at this plan shows that the placement of the hinge and framework required digging into the socle two grooves in the form of steps, large enough in every direction to place these pieces of wood, and this was done.

With the monolith thus framed, all we had to do was start off the first push, which would support the force of the mass left to its own weight.

Lowering Equipment, Plate II

This first piece of equipment was composed of three capstans that required the same number of pulley blocks. The pulleys of these blocks were lashed on one side to the far end of the doubled cable under the pyramidion, and at the other end to anchors in the sand. Each capstan was turned by sixty-four men, thus pulling toward itself the lowering cable and consequently the head of the monolith, so that the latter, yielding to the combined actions of these machines, had to lean toward them in pivoting around the hinge.

Having arrived at this position where the mass would tend naturally to descend on the roller, far from helping this motion by traction, it had to be restrained with a second apparatus to avoid unwanted acceleration in the movement of rotation.

Fig. 1.

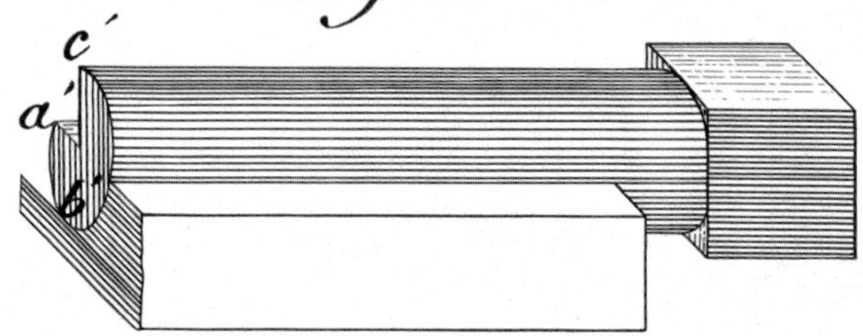

Detail from Plate II: The hinge around which the obelisk turned

To understand properly the maneuver that this apparatus carried out, it is necessary to consult the details reported on page 54 and to assume that all the fill (rubble) U between the obelisk and the roller had been removed up to the level of the base. In other words, we created a vast excavation—BCOU—following the axis of the road, on one side by a vertical plane passing through the lower east edge of the socle, and the other by a plane UO parallel and tangent to the roller.

Retaining Apparatus, Plate II

We constructed on the bottom of this excavation BCO a platform, NN, composed of several pieces of crossed wood that ended at plane UO.* A long timber or plank,

* See appendix for the explanation of the technical terms in this section.

PP, pegged to the extremities of this frame and also held back by strong wedges, served as a stopping point for the apparatus. The brick wall that rises above was meant to contain spoil from the debris. It is on this platform that the main part of the system of retainers rested, the support for the shear legs.

The support had as its base a strong piece of wood that ended in a half cylinder and was placed in the bottom of the right angle NPD formed by the vertical plane of brick wall and the support platform. On its flat surfaces, whose middle corresponded to the axis of the road, were tenoned eight masts or pulleys divided into two symmetrical groups of four each, one going to the right and one to the left of the obelisk. These masts were connected at the top by a double crossbeam forming a large trapezoid that could turn around its base as well as around its hinge. The double traverse beam was tied to the top of the monument by cables or rope passing and tied under the pyramidion. In order to better guide the monument in its descent, a system of ropes was deployed in a fan and moored to the end of the pulleys. Each one of these carried at its little tip a pulley* of three spools. These pulleys corresponded to an equal number of two-spooled pulleys† tied into a grillwork that encircled the eastern pedestal to form, along with the eight masts to which the retaining pulleys were attached at the same fixed point, a complete system.

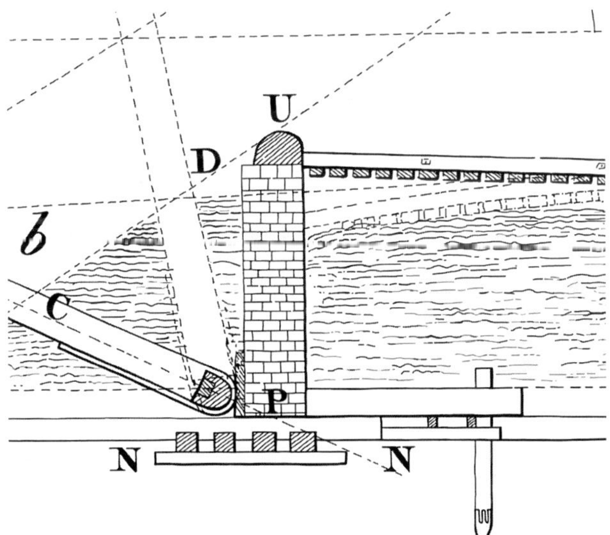

Detail from Plate II: Considerable rubble had to be excavated to secure the apparatus for lowering the obelisk

Each winch was furnished with six branches.** The major tackles rolled around each one in a double turn,‡ first on the groove of the winch TT, horizontal and free to turn on its axle, then on a fixed cylinder MM, where they would slide with friction, passing from there into a return pulley, V, then come back a last time with friction on a yard arm VV, and end up in the hands of eight men. Such was the whole retaining system, of which only part of the hardware has been described.

* See appendix.

† See appendix.

** See appendix.

‡ See appendix.

We now need to describe the working of the apparatus, to enter into several details of its various parts, to examine the function of each component, and finally to calculate the forces exerted on the fixed points, the tensions on stays and the apparatus going out and coming back.

At the start of the movement, the obelisk is vertical, the derrick is in BB, and capstans pull in the direction FH. They are manned by 192 Arabs; the monolith, yielding to the action of these engines, will lean toward the bank, dragging with it the frame of the derrick, to which it is tied in a fixed manner by a system of cords deployed in a fan. The latter will act in turn on the retaining winches, causing them to unwind. Because of the friction and stiffness of the ropes, combined with the action exerted by eight men on the heads of the apparatus, we planned to adjust at will the movement of the pulley block which controls the movement of the mass. For the entire rotation, the obelisk will rest on the pedestal. A part of the weight would thus be carried by this resting point. The other part, which constitutes the resistance to be overcome, obviously depends on the angle at which the monolith will be pulled by its cable, FH.

According to these arrangements, the tension transmitted to this cable by the actions of the capstans will be 28,518 kilograms,* a force that exceeds the maximum resistance of 5,000 kilograms. But this excess of force is necessary to overcome the friction and stiffness of the ropes of the retaining apparatus, whose heads will be spun out little by little as needed by the mechanism.

Having arrived at Position C, where the vertical of the center of gravity goes beyond the base, the obelisk will fall over on its own, and the role of the capstans will be finished. The mass, thanks to the accelerative force of the weight, will continue to turn, exerting on the top of the shears, and through them on the derricks, a constantly increasing force, since a larger part of its weight will fall beyond the pedestal. This power, varying with the inclination of the mass, will pull successively, from bottom to top, the mobile pulley of each retaining pulley block with a force of 4,000 kg, 7,080 kg, 9,500 kg, or 11,880 kg at the moment when the center of gravity corresponds to each point marked on the plan in the appendix by the numbers (1), (2), (3), (4). The last one will correspond to the last part of the fall.

These pullings, enormous relative to the strength of a man, will be absorbed for the most part by the friction and stiffness of the ropes. The apparatus is put together in such a way that they will only transmit to the extremity at the top a corresponding tension equal to 4.34 kg, 7.67 kg, 10.30 kg, or 12.88 kg. Thus, in all of the intermediary positions between points c and (4) the force expended on each of the cords to equalize the force, and to reestablish the immobility of the system if necessary, will be less than 12.88 kg. Since a robust man can produce a force of 20 kg, it follows that the eight sailors

* See appendix.

controlling the movement will be able to accelerate, slow, and even stop the fall of the obelisk, if need be.

These results, deduced theoretically (see appendix), assume an equal tension in all of the pulleys. But it is not so in practice, whether you need to raise a weight or desire to lower it to the ground, with the help of many derricks. Since the chief cables of the apparatus are not being let out by the same hand, it is physically impossible to exert upon them a uniform movement, resulting in an uneven distribution of the weight on the shears. It is in order to avoid this serious problem that we use, in big jobs like this, pulley blocks whose absolute force is double, sometimes triple, the resistance to be overcome. The system of pulleys of the apparatus of Luxor was far from offering this excess of strength; the strength of the sun had caused cracks or splits in the encasements of the pulleys, one of the derricks broke under a force of 13,000 kg when tested, and we came to realize that they had to support a tension of 12,000 kg, supposing equal tension.

This uniformity of tension was one of the necessary conditions of the problem, and all the more difficult to fulfill because we were missing the necessary materials. This necessity suggested to me the idea of regularizing the movement of the apparatus by means of a mobile cylinder TT. The winch, which at first appeared useless, was a very great help in the most delicate moment of the operation. It bore, as we have said, eight grooves of the same diameter. They were formed by brackets attached to the pole to make them rough. The main cables of these tackles were wound around these grooves, each by a double turn. The resulting friction made sliding impossible. Thus it was only through the revolution of the winch that the pulley blocks could unwind. Because this machine was unwinding each apparatus equally, the mobile pulleys had to move away from the fixed pulleys at the same time and at the same rate. All that was needed was to give to the system of shear legs a uniform tension before starting the motion (which is easy with a little practice), to make sure that they would maintain it during the entirety of the rotation.

From the equilibrium equations (see appendix) we deduced the successive tensions of the shrouds and the tackle and the pressures on the turning pivot and the foot of the shears. Taking into account only the worst-case scenario, we see: (1) that each shroud will have a tension of 6,416 kg against a breaking strength equivalent to 19,980 kg; (2) that the most highly stressed cordage will have to resist a pull of 2,910 kg, which is less than the force of 3,252 kg that they can support without breaking; and (3) that the point of application of the moving pulleys will be stressed vertically by a force calculated at 71,000 kg with which to oppose a resistance of 80,000 kg, not counting the friction of the gratings on the eastern obelisk.

Finally, the horizontal components of the pressures exerted against the pedestal and on the stop at the foot of the shears, being a reactive force, are partly dissipated;

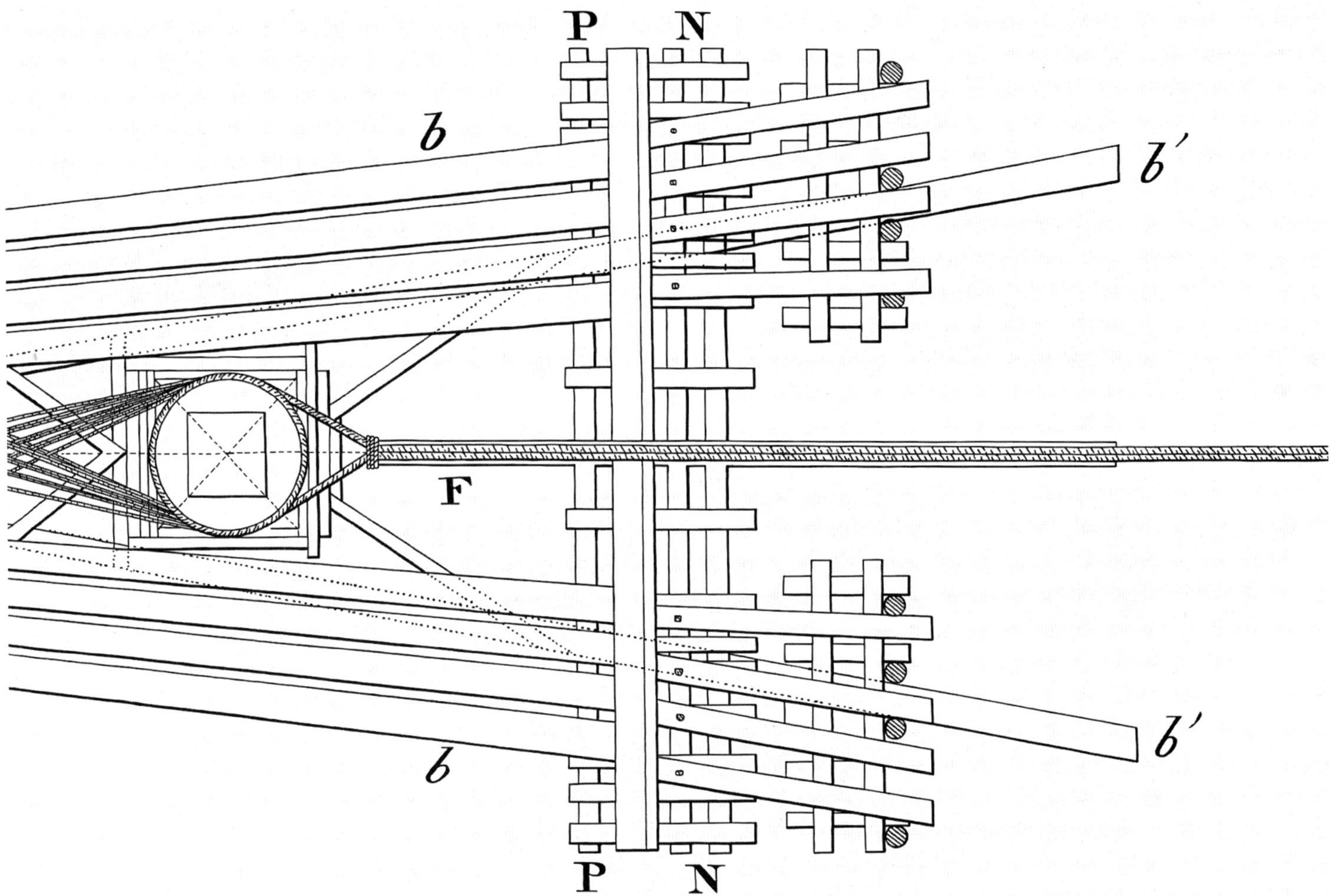

Detail from Plate II: The platform (NN) built to stabilize the lowering apparatus

the excess of the first over the second is neutralized by the mass of sand piled up against the east face of the socle and by the friction resulting from the platform, NN, to which this block is tied in a rigid manner by means of the horizontal shears, b′b′.

Thus, all the elements of the system meet the conditions of forces and resistances necessary for the success of the maneuver. The only difficulty that remains to overcome consists of constructing the apparatus as we described before, with the material and number of workers. As for finding right here the wood and men capable of working them, it was better not to think about that.

I had brought from Toulon, according to my instructions, eight carpenters and two blacksmiths. These workers, under the orders of a skillful master, and helped by the sailors, got to work as early as the first days of our arrival at Luxor. The carpentry work was necessarily executed in the place where the obelisks stood, in the middle of the whirlwinds raised by the four hundred laborers digging the roads. This sand, divided into the smallest particles, penetrated to every part of the body, mixing with the sweat, creating a general smarting sensation, sticking in the throat, oppressing the lungs, and provoking coughing and great thirst. Anyone who has not lived in Upper Egypt can have no idea of these painful manifestations. Despite all these sufferings, we still had to spend the day with the hatchet or the adze and supplement the workers with one's own activity. But the idea of contributing something to an operation whose purpose was to endow France with a remarkable monument was sufficient to sustain the energy and the indefatigable zeal of the French workers. From daybreak, one could see them going happily about their tasks that, in other circumstances, would have seemed rough, difficult, and dangerous to them.

There was no end to the suffering and the obstacles, but a much greater calamity was coming to summon up a new test of their patience and devotion. On August 27, some sailors on the Nile arrived and announced to the inhabitants of Luxor that a plague of cholera was spreading its horrible ravages throughout Lower Egypt. Several days later, the quays of the river at Thebes were filled with boats loaded with Europeans fleeing the epidemic. Everyone we encountered had stories of what they had seen at Cairo or Alexandria, and all agreed that the malady was rapidly on a march toward the Said (south).

On September 10, this sad news was confirmed by the sudden death of a Turk who had arrived the previous day from Gurna.* The tragic end of this unfortunate fellow left no possible doubt about the nature of the pestilence menacing us. Cholera was at our door. The next day, it penetrated the town, then the work site. Every day we lost more workmen, some of them even falling at the very moment when I was distributing their daily

* Village situated on the left bank of the river, opposite Luxor.

wages to them. This deplorable situation, and the tears and lamentation of the inhabitants, made a terrible first impression on us, but our great task allowed us no time to reflect on the perils that surrounded us. Nobody quit his post; no work was suspended. One single idea animated us all: to remove from the ancient capital of the civilized world one of its most beautiful ornaments.

However, our situation became more and more critical and a number of circumstances came together to aggravate it. Fifteen sailors came under the dreadful attack of the cholera, and our communications with Alexandria were interrupted. The materials that we were awaiting from that port to construct the haulage road would no longer be arriving. Left on our own, devastated to have nothing to hope for from outside, we knew we would have to supply anything we needed from our own efforts. Time was pressing, the work was already seriously behind schedule, and the number of workers available was diminishing every day. Everything contributed to my fears that we would not be able to complete our operations before the next flood. The lack of any correspondence with the mother country, or even with Alexandria, made us very nervous, and our low morale thus aggravated our physical privations. To sustain the courage of our compatriots and to keep them from disregarding the regimens prescribed by the doctor of our ship, we employed them only in special projects, using Arabs instead whenever possible. We had to do this, given our disastrous position.

My first care was to organize the long (ripping) sawyers, as our own workers could not have stood up to such a difficult job even in the south of France. But four Arabs, directed by a carpenter, after having practiced several times on the trunks of date palms, acquitted themselves well enough to start sawing out the oak planks that we needed. Six months later they were able to saw with precision, not only straight lumber, but also pieces with double curvatures. A squad composed of thirty workers and twenty apprentices took over all the carpentry work. The former, who in the beginning, scarcely dared to touch the planks put on their shoulders, and who had no conception of trailers, carts, sounders, capstans, etc., now could use all these instruments with military precision. In a few days, the children learned the names of all the tools and how they were used. They helped the carpenters, taking measurements, planing the wood, etc. They went to the arsenal to find various objects needed so that the craftsmen would lose no time. Similar practical results, obtained after a year of experience, sufficiently proved the intelligence of the Arabs—and I must acknowledge that they were of great help under the circumstances.

Although the epidemic had slowed our work activities, as early as the first of September the obelisk was cloaked in a light casement made in part, for want of other materials, from the bars of the capstan. The pulley blocks strapped to its top lifted the boards destined to cover its faces. These results obtained amidst the rubble

Luxor Temple as drawn by Verniac Saint-Maur, captain of the *Luxor*

were much better than we could have reasonably hoped, especially considering that sickness had recently decimated the population of the village. The huge excavation that I noted earlier was completely finished. One could now walk all around the obelisks and examine all their faces while walking on the pavement of the square in front of the edifice that for many previous centuries had lain hidden beneath the desert sands. By imagining that the ruins that still remained on the sand had been cleared, I could see the façade of the temple, not as it is today with its broken architectural structures, heaped up, buried in mounded piles, but as it must have looked in the era when the best of science and wisest of men combined to produce it, when their artistry was spread out in all its magnificence.

At first, the eye has trouble accommodating to a spectacle so new, so unexpected—an architecture that resembled nothing that we had seen and admired before. Here there were no statues with missing members, no doors, no windows, no volutes of twisted leaves as great as in the days of Greece, even less of elegantly supported architraves on colonnades on four ranges of arches with a triple volute in the ribbing; here were only colossal masses placed before yet more colossal masses.

These are twenty-four-meter-high obelisks covered with carvings, colossi whose calm and immobility, whose regularity of proportions, command respect. Two pyramidal masses between which one finds a door seventeen meters high, crowned by a half-arch and surmounted by an elegant crown, make an ensemble that clearly portrays its characteristics of durability and simplicity, of grandeur and excellence, of execution that cannot be found to the same degree in any modern building. This immense palace, sculpted and painted, outside and inside, with the most vivid colors, had to be one of the marvels of Thebes. Even as it is today, it gives an idea of the much grander form of its antique magnificence.

The square of this palace is composed of two levels of large crossed flagstones. At 1.2 meters from the façade are seated four monolithic colossi; a bit more forward are two pedestals on which the obelisks rest.

The pedestal of each obelisk is composed of two distinct parts, the base and the dado. The base, formed by the union of three blocks of silica sandstone, resting on the first level of the flagstone, is embedded into the second. The dado is a granite monolith.

The south side of the west dado bears cynocephalic [dog-headed] apes. The first three are sculpted within the cube itself; the fourth is carried over to the east face with half its thickness. This one and its adjacent side, its northern one, are neither flat nor vertical, whereas the west side presents a tableau of perfect hieroglyphs. These irregularities indicated that the dado had not survived complete. We found, indeed, in the excavation debris fragments of a granite plaque that used to cover the east side, the remains of four cynocephalic apes standing against a square of the same dimensions as the north face, in such a way that the base in its earliest state must

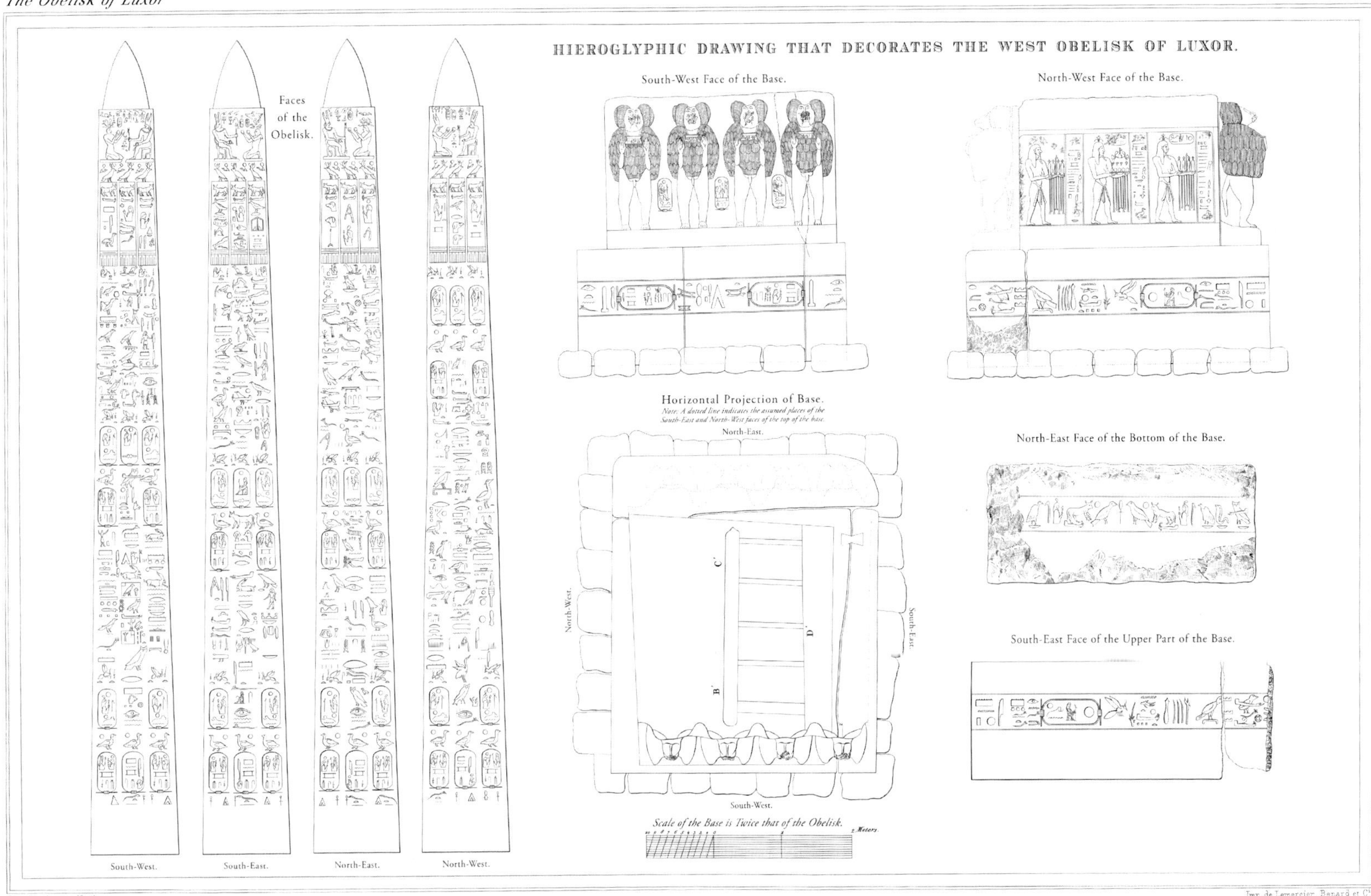

Plate III: Hieroglyphic drawing that decorates the west obelisk of Luxor

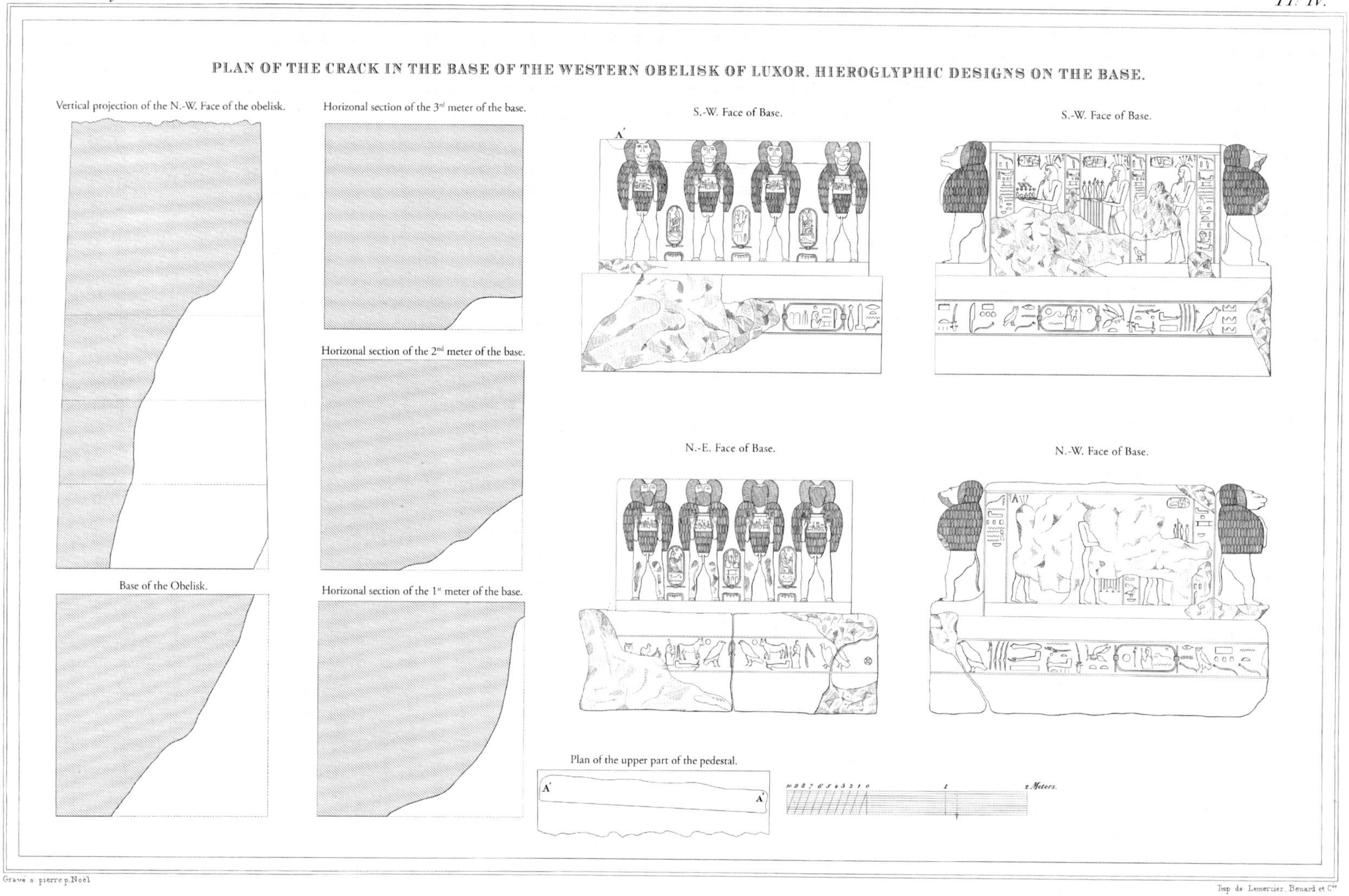

Plate IV: Plan of the crack in the base of the western obelisk of Luxor; hieroglyphic designs on the base

have had the form shown in Plate III. The east pedestal, uncovered later, is as we have drawn it on Plate IV. It is in all aspects similar to the other except that the four cynocephalic apes are of one piece with the dado. The smallness of the block used in the western base did not permit the sculpting of those four figures. The artist supplemented this with a built-up piece and a slab.

The height of the pedestal at the right is 3.4 meters, that of the pedestal on the left, 2.6 meters. This inequality is naturally explained when one notices that the two obelisks, which are pretty much facing each other, do not have the same dimensions. In order to remedy as much as possible the disadvantage of the disparity, the architect conceived of placing the smaller (the Paris one) on a base higher and a little more forward from the temple. By this positioning, he distorted its proportions by bringing it forward to the eye of the spectator. He also had to hide the defects of the pyramidion with a revetment in order to restore it to the desired proportions. The seven to eight-centimeter ledge that runs around the base would seem designed to support it, but this restoration, like that of the dados of the base, could only be done with granite. Any other material would have created a marked difference between the two obelisks because it would have been impossible to adopt a similar coating on the left pyramidion. This one here is perfectly polished and its faces are flush with the shaft.

As I said above, as long as I did not have a precise, definitive idea on the system of work, as long as the work still remained to be completed, I was completely dominated by the thought of the task I had to accomplish. I saw in the monument nothing but the object of my task, a block of granite, a stone, as Krali Effendi called it. Now that I am pretty certain to take it down from its base without incident, now that its face is cleaned of the shroudings that concealed it and its base is cleared of the age-old sands that hid it from the eyes, allowing us to observe the mysterious inscriptions carved on its shaft, we must, I believe, go into historical details of these imposing masses.

Obelisks are of one single stone, pink granite, placed always two by two at the entrance to buildings, forming an integral part of the buildings and linked to a whole system of ideas. They were conceived and erected with the idea of preserving for posterity the reason for the creation of the monument they decorate: the name of the divinity to which they were dedicated and the honorary titles of the sovereign who had them raised. They are positioned on low bases in order to emphasize the height of the shaft, lit by streams of light that accentuate the color, the shine and polish of the granite, and the purity and fineness of the hieroglyphs. These monoliths whose summits are outlined in the sky produce a magnificent effect of perfection that augments the grandeur of their proportions. This illusion was so complete that it equaled, in my mind, the very demonstrations of geometry. Looking at these masses, I could not help but feel a vague sentiment of malaise and uncertainty.

However, the means that I had put together to manage to move them were based on sure calculations that I truly believed guaranteed the results.

One finds nothing in Egypt more perfect than the three columns of hieroglyphs that decorate each of the sides of these monoliths. They start about a meter from the bottom and they end at the pyramidion. The middle column is incised deeper than the two others. It is worked and polished with as much care as fine stones; it is especially remarkable for the sharpness of the edges of the signs sculpted there and by the excellence of the details. The outer columns are only broadly outlined. Because of this feature, the rays of light, reflected differently, produce different shades that prevent all confusion and permit us to distinguish clearly this monumental writing.

Part of these hieroglyphs could not be seen because the bases were buried under the debris, and one side of each of the monoliths was partly hidden by modern construction. It was only after the completed work of which we just spoke that the figures were uncovered. I hurried to send the drawing of them to the minister of the navy and Monsieur Champollion the Younger.

We could not do any better to satisfy the curiosity of the reader than to take an extract from the following scientific work published by Mr. Champollion-Figeac, from the notes and interpretations that were left by his illustrious brother.

Inscriptions on the Obelisk

The four faces of the obelisk are carved with inscriptions in hieroglyphic characters.* A superficial examination is sufficient to show that on each of them the signs are placed symmetrically to form three perpendicular columns, clearly distinct, and forming thus three inscriptions, three sentences on each side. Each side can be divided into three parts with regard to the entire inscription it contains, as follows:

1. Immediately below the pyramidion, bas-reliefs of offerings that occupy the entire width of each side.

* The Egyptians used three kinds of characters for writing. For monuments: *hieroglyphic* signs, where each is the exact representation of a plant, a quadruped, a bird, or any object of creation or of human industry. For books and other subjects relative to religion: the priestly *hieratic* signs, simple shorthand hieroglyphic writing. For contracts and daily life: the *demotic* or popular signs, simple abbreviations of hieratic characters.

2. At the top of each column of hieroglyphs is the figure of a symbolic sparrow hawk, coiffed by the whole crown *(pshent)*, on top of a frame that ends in fringes in its lower part. One can give to this framing the name *royal banner*. It contains the varied honorific titles of the princes named on the obelisks, and it is found, on its own, next to the Egyptian kings, in their representations of religious or civil ceremonies.
3. The inscription itself, whose signs, divided into three parallel columns and written one below the other on their own or in groups, form three vertical inscriptions that are read from top to bottom.

The Paris obelisk recalls the names and actions of two kings together, but history, in all its fairness, can give to each the part he deserves. It was Ramesses II who had the obelisk extracted from the quarries of Syene, who had it transported to Thebes, and who destined it to decorate a great edifice. The identity of this building is difficult to figure out today. It was not the palace of Luxor. The name of Ramesses II is not found on any part of this palace. Could it be, with its foundations, still hidden in the sand? If so, might it be necessary to add a second supposition, attributing to this king the building of the main parts of the Rameseum of Luxor, which Sesostris would have decorated and completed?* But all the possible conjectures are equally uncertain, each one able to gather in its favor a sum of almost equal evidence.

It is certain that the purpose of the two obelisks was to transmit to us the memory of the glory and piety of Ramesses II through eight inscriptions. Only four of the inscriptions were completed. Three are on the obelisk transported to Paris and one on the obelisk that remains at Luxor. What happened to interrupt these chants of victory? Death took Ramesses II in the middle of his trophies; his victories over the infidels (foreigners) are indeed mentioned there. The obelisks are later than his campaigns in Asia and Africa and can only belong to the end of his reign, which was of rather short duration.†

* Translators' note: Champollion was wrong about the Luxor obelisks containing the names of two kings. Only the names of Ramesses II (Ramesses the Great) appear on the obelisks. All pharaohs had five names, and Champollion mistakes one of these names for that of another king, whom he calls both Sesostris and Ramesses III.

† Translators' note: Ramesses II actually ruled for about sixty-seven years.

Then came Sesostris, who built or completed the Ramesseum of Luxor; he adopted the obelisks that had been started by his predecessor, and he used all the space that the death of Ramesses left unoccupied to recall his glory. That is to say, he inscribed three entire faces of the obelisk still in Luxor and an entire side of the Paris obelisk, and on each of the three other finished sides, as well as on the only one occupied by the name of Ramesses on the other, he put two side inscriptions, for which there was enough space on all of the faces.

On the obelisk of Paris, the work of the two kings are thus distributed:

On the north, south, and east faces:

Ramesses II, middle inscription

Ramesses III, the two side inscriptions

On the west face:

Ramesses III, all three inscriptions

Further, Ramesses III had this obelisk raised and his name engraved on the top plane of the base, and on all the parts of the pedestal where this name could be placed as an ornament or as a historical document.

Finally, and in order to multiply again this information for posterity which would extend until our present generation, and which it was part of the destiny of France to perpetuate by its munificence, Sesostris had written on the north face of the obelisk left in Luxor that he himself, Lord of Upper and Lower Egypt, Son of the Gods and Goddesses, Lord of the World, Sun Guardian of the truth, Approved by Phre, has made these works (the Ramesseum of Luxor) for his father Amun-Ra, and he erected these two great obelisks in stone in front of the Ramesseum in the city of Amun (Thebes).

Sesostris thus finished this grand work begun by his predecessor, and the confluence of these two kings for the achievement of these admirable monuments has furnished somewhat precise chronological information on their history.

The reign of Ramesses II, who began these obelisks, goes back to the year 1580 BC. There exist no monuments with dates after the fourteenth year of his reign, which ended soon after that. Thus, it was around the year 1570 that these obelisks were undertaken by Ramesses II after he punished the infidels in Africa and Asia, as his inscriptions tell it.

Sesostris succeeded his brother around 1565. He built or continued the palace of Luxor, a work that demanded many years. On the bas-reliefs of the pylon, which is the very frontispiece of the palace, Sesostris had sculpted in large format his campaign against the Asiatics, and the inscription gives it the date of the fifth year of the reign of this king. The obelisks were not raised until after this pylon. One can thus believe that they were there as early as the year 1550, where for 3,400 years they have braved both time and man.

Text of Hieroglyphic Inscriptions

The translations and summaries that follow are taken from the manuscripts of Champollion the Younger.* We have limited our extracts to those that will be of particular interest to readers, and in order to follow the order of importance of the hieroglyphic inscriptions engraved successively on the obelisk, we will first present those concerning Ramesses II, predecessor of Sesostris.

Inscriptions of Ramesses II, Armais

West face. Side facing the Arch (of Triumph) in the Place de l'Étoile (north face at Luxor)

Bas-relief of offerings: The god of Thebes, Amun-Ra, is seated on his throne. Two long feathers ornament his headdress. He holds in his right hand his ordinary scepter, and in his left the ankh, symbol of divine life. In front of him, Ramesses II is on his knees, his head adorned with the headdress of the god Ptah-Seker, surmounted with a winged globe. He is offering to the god Amun-Ra two flasks of wine. The cartouches, names, and prenomens of Ramesses II are in front of his image, and the legends of Amun between the cartouche and the headdress of the god. The short inscription, perpendicular to his scepter, is also the title of the tableau: "Gift of wine to Amun-Ra."

Middle column:

a. The Banner of Horus, "Powerful, beloved of Satis."
b. Vertical inscription: "Lord of Upper and Lower Egypt, Regulator, Lord of Egypt who has punished the countries. Horus, resplendent, guardian of years, great of victories,

* Translators' note: Champollion died on March 4, 1832 while Lebas was still in Egypt. Champollion never published his translation of the Luxor obelisks.

king of obedient people, SUN GUARDIAN OF THE TRUTH, moderator of moderators, engendered by Atum in . . . with him, to exercise the royal powers over the world a great number of days. . . . The city of Amun (Thebes) . . . Son of the sun. THE BELOVED OF AMUN, RAMESSES. May he live."

East face. Side facing the Tuileries (south face at Luxor)
Bas-reliefs of offerings: The subject of this bas-relief is analogous to the north face. Ramesses II makes a new offering of wine to the god Amun-Ra. The banner also contains honorific expressions of the king. The inscription also contains the praises of this prince, his same names and prenomens, the titles of "Resplendent god, Guardian of the vigilant ones," and an invocation to Amun, who has here the rank of Lord of the Gods. It reminds us that the king has dedicated a sanctuary consecrated to a divinity, and at the same time he has honored the other gods of the same temple. These circumstances allow us to assume that Ramesses II had intended the obelisks in Luxor for a temple other than the one in this place.

North Face. Side facing the Madeleine (north face at Luxor)
Bas-reliefs of offerings: The same king makes the same offering to the same god. The vulture, emblem of victory, flies above the head of the king.

The middle column of this side is also that of Ramesses II. The banner still bears his royal or religious titles. The same goes for the inscription that gives him the titles "Guardian, Great of Conquerors on the entire earth, visible sun, etc." The inscription ends with the name of the king, and the wish, "Forever."

Inscription of Ramesses III Sesostris
West face. Side facing the Arc de l'Étoile (east face at Luxor)
The bas-reliefs of the offering pertain to Ramesses II.

Left column. Banner: "Horus powerful, guardian of the vigilant (or overseers)." The inscription recalls the strength and victories of Sesostris and his glory in the entire world.

Right column: The banner describes him as the beloved of Satis (Truth). The inscription says that the entire world has trembled because of his exploits. It assimilates him to the god Montu, whose son he is said to be.

South face. Side facing the Pont de la Concorde (west face at Luxor)
This side of the obelisk belongs entirely to Ramesses Sesostris. Ramesses II had left it empty.

In the bas-relief of offerings, Sesostris, topped with the complete *pshent* (crown), symbol of his authority over Upper and Lower Egypt, and surrounded by the winged globe of the sun, offers wine to the great god Amun-Ra, the namesake of Thebes.

To the customary praises, the middle column adds that Sesostris is "the preferred son of the king of the gods, the one who is on his throne, rules over the entire world." The palace that he erected in the south (i.e., the southern part of Thebes) is mentioned. The title "Beneficent" is given to him in the inscription on the right, which adds, "Your name is as stable as the sky; the duration of your life is equal to the duration of the solar disk." In the inscription on the banner on the left, Sesostris bears the title "Beloved of the goddess Satis," along with other praises very common in the Egyptian royal protocol. This inscription declares Ramesses III the son of the king of the gods, to take possession of the entire world. The three columns of this face are all terminated by a cartouche of the proper name of the king, "Son of the sun, beloved of Amun, Ramesses."

East face. Side facing the Tuileries (south face at Luxor)
The banner and inscriptions on the column on the right proclaims Sesostris "Horus powerful, friend of truth (Satis), moderating king, most kind like Atum, being a chief born of Amun, and his name being the most illustrious of all."

On the column on the left the banner reads "Horus, the living king of the Upper and Lower Regions, child of Amun." The inscription gives to Sesostris the title of chief king, mentions his works, and adds that "he is great through his victories, preferred son of the sun in his royal abode. The king (his prenomen and proper name), the one who makes Thebes rejoice like the firmament of the sky through his considerable works, forever."

North face. Side facing the Madeleine (east face at Luxor)

The banner of the left column is noteworthy for the great number of signs composing its legends. It says: "Horus, the powerful, the great one of conquerors, fighting on his strength." The inscription names Sesostris "the great conqueror, lord of victories, who led the entire country and who is most kind." Finally, the banner that surmounts the inscription on the right announces that Sesostris is "Horus the strong, powerful in the great panegyrics (civil or religious assemblies), friend of the world and just king." The inscription adds, as if to heap up the measure of praises, that he is also "prince of the great ones, enjoying royal power like Atum, and the chiefs of the inhabitants of the entire world are under his sandals."

Chapter 5

Lowering the obelisk and loading it onto the *Luxor*

After this brief excursion into the domain of history following the footsteps of the famous archaeologist, we must return to our work.

On October 1, 1831 the obelisk is protected by its covering, the base of the scaffold is about to be put into place, and the masts suspended from the pulley blocks are ready to go into their mortise. There remain only a few essential details to take care of, for which, to tell the truth, we lack some materials, but these difficulties will be solved and everything makes us hopeful that before the end of October the monolith will be taken down from its base.

With the scaffold of shears set as we indicate on Plate II, we proceed to the installation of the shrouds. Each of these cables, placed separately and looped under the tip of the obelisk, is tied to the extremity of the shears, after going through the eye or grommet of a pulley with three wheels. Before the final tying, the whole system is uniformly tense. The lowering cable is rigged at its end, a wooden half-crescent where the mobile pulleys of the control derricks are attached. All these maneuvers were executed by sailors suspended in the air.

Lacking the necessary wood for the construction of the supports of the winch, we used two capstans, C′C′, temporarily for this purpose.

On October 23, the pulley blocks and derricks are fitted with their tackles. The scaffolding that surrounds the obelisk is removed, and the sailors affix the tricolor flag on top of the sheers, surmounted with garlands of flowers and palm branches. Everything is finally ready for the lowering of the monolith.

Dawn found everyone at his post. One hundred ninety Arabs, scattered on the bars of the capstans, awaited the agreed-upon signal to start the motion.

The eight topmen were there, holding in their hands the main cables of the retaining apparatus. In a few minutes these men alone would maneuver an immense load, control its rotation, and resist its immobility, as needed, at any given point of its course. They were proud of the important role with which they were entrusted. The greatest anxiety was seen on all the faces; when the sun touched with its first rays the colossal statue of Memnon, which at one time, animated by the celestial flame, marked the air with harmonious sounds, this was the signal for action.

The voice of the commander was heard. Right away the capstans turn on their axes. The ropes of the derricks stretch, the tension increases. It accumulates, reacts on top of the shears, and through them on the retaining pulley blocks. The resistance and the friction are dispelled; the winch, activated by the power, oscillates around its axis; then it rides on its pivot, and the obelisk, finally detached from its ancient base, slowly inclines its pyramidal head toward the riverbank, taking with it the scaffold of the shears, which simultaneously turns around its base. After going through an arc of eight degrees, it stops and the whole system again becomes immobile. The officer in charge of the capstans announces that the point of support of the derrick broke, that the anchors are dragging through the sand. This movement indicates that the retaining cables are too tense and the cable needs to be played out faster. This maneuver is executed with precision, the motion continues, the rotation starts again. We see the lower edge climb gradually, and soon the vertical of the center of gravity has passed the hinge. From this instant, the work of the capstan has to stop. The difficulty is no longer the inclining of the obelisk; now it needs to be supported with retaining cables so that, once it gets going, it doesn't throw itself roughly on the roller. The eight top men now start to exert successive degrees of tension on the main cables proportional to the length of the one arm of the lever, which always increases as the vertical of the center of gravity gets further away from the pedestal.

At a given signal the monolith stops and remains suspended for two minutes at a twenty-five-degree angle. At a second signal it continues its progress with the same order and the same measure until the west side reaches the cylinder set in the brick wall.

The first rotation was done; the obelisk had just described an arc of about sixty degrees (see Plate II).

This maneuver lasted twenty-five minutes; workers and sailors had all accomplished their duties. All the orders had been executed with the greatest precision and a remarkable intelligence. Thanks to their zeal and precision, no serious accident saddened such a beautiful day.

Careful examination of the cladding and the predicted trajectory confirmed that nothing had moved. There was no visible deterioration of the pivot of the obelisk, which had supported for twenty-five minutes a pressure of more than 250 tons. This experience gives an idea of the strength of oak wood to resist crushing.

If I had any doubts left regarding the preservation of the monolith, the inspection of the base alone would have been enough to dissipate them. It was covered with a slight layer of silt that did not show any continuity. At first glance, it looked as if the groove was not in the granite, but after removing the silt that had been hiding the surface, the groove presented itself as shown at P' (Plate II).

There existed in the crack *mn*, which joins the two longitudinal troughs, a resin that did not stick to the stones very strongly. It had not undergone any degradation. This same cement was found around the two mortises dug in *m* and *n*. These cavities were filled with a yellowish dust

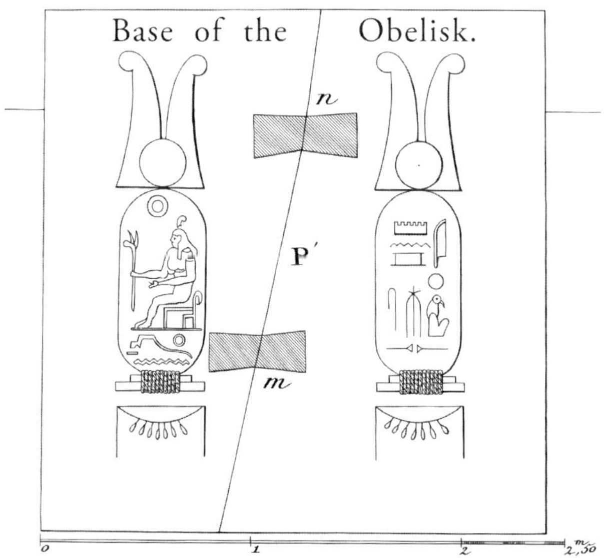

Detail from Plate II: The crack running through the pedestal on which the obelisk rested

mixed with sand and silt that came from the remains of wood dovetails that had been inserted to stop the separation that the crack could have caused. We determined that this was a fault in the stone as ancient as the monolith itself, whose antiquity is attested by the names and prenomens of Ramesses sculpted under the base.

The Egyptians frequently used this kind of binding; a great number of flagstones that cover the temples of Luxor and Karnak are joined two by two in the same way. The dados (cubes) of the pedestals show some similar cuts and are filled with granite keys. They were used to affix the cladding with which they were covered. Grooves were also visible on their upper planes, A′B′C′D′; the one dug on the eastern block at A′ particularly captured my attention. It is semicircular; its axis coincides with the lower edge of the monolith. Since the erection of this monument preceded that of the western obelisk, it was possible to take it from the banks of the river up to the top of the pedestal in such a way as to have those two lines join. This circumstance permits us to believe that the groove that we just spoke of was made for the purpose of putting inside it a cylinder that would serve as an axis of rotation when the obelisk was raised on its base.

It seemed quite strange that after more than thirty centuries, the same type of operation would be used to move these masses.

The success that had just crowned our first efforts had been bought by hard work, care, and ingenuity;

however, the following operations offered still more difficult challenges, because of our limited means. One cannot imagine how much difficulties cost in a country where one cannot find the materials.

Before proceeding to the second rotation that was to leave the obelisk on the towpath, it was necessary to prepare a wooden casing to receive it. Without this precaution it would have been impossible to slide it on sandy and freshly dug-up ground. Here we were lacking necessary materials for the construction of a slipway, which would have started from the roller and ended at the front of the ship. The platform NN would have used most of our wood, although we could have used palm trees for all the parts that didn't require nails or bolts. Its first plane was entirely composed of pieces of this soft tree, which has no resistance and is good only to serve as filler. We had not yet received any answer to the request for planks addressed to Alexandria in July and requested again in August and September. Waiting longer would have meant losing precious time. Later, with the winds from the south not allowing us to work, the inundation would have reached us before we could have loaded the monolith. These considerations were determining factors. Using the rest of our supplies, which by now amounted only to several pieces of boards and six beams of eighteen to twenty square centimeters by seven meters long, we had to come up with a system that could transport the monolith a distance of about four hundred meters.

This slipway, twenty-one meters in length, was divided into lengths that could be taken apart. Each of these parts, which we will call flooring, was composed of two parallel beams. They rested on a floor formed of pieces of planking placed across and nailed under their lower sides. These floorings, placed end to end, were tied together by the round pieces TT bolted at the two extremities, which allowed us to move them and to carry the flooring from the back to the front.

The small amount of space that the wooden construction occupied on an unstable ground, the small number of important parts, the space of fifteen centimeters left between every two boards in order to economize on wood, all tended to diminish its strength. In a word, this contraption did not offer the amount of resistance necessary to haul such a great mass. We could therefore expect major shortcomings and suffer all the consequences.* Once this work was completed, not without difficulties, and the slipway installed at UT, the master carpenter came to let us know that all of our materials had been exhausted.

"I have nothing left," he told me, "not even a piece of wood to make a tongue or bracket, which we will probably need during the operation. If, as you fear, the planks should break, it will be impossible for me to fix them."

* It was these considerations that made me decide to send the report about taking down the obelisk only after the completion of the operation, whose success at this point was, at the very least, doubtful.

These observations were perfectly correct, but before worrying about the future, we had to provide for the important necessities of the present, at least as much as our resources allowed it.

Second Rotation of the Obelisk

We had left the obelisk in position XY. Plate II indicates the way it was placed on the towpath.

The center of gravity of the mass being a little toward the front of the roller, the monolith, left to its own weight, was supposed to flip of its own accord around its axis; but in this rotation, as in the first one, it was crucial to control the speed in order to avoid a jolt at the moment the mass arrived on the prop.

The primitive apparatus, stopped in its motion by the brick wall UP, could not be used for this purpose any more. The operation required us to untie the ropes that attached the tip of the obelisk to the tie-beam of the scaffold. When this was done, the same cables were used to maintain all the rigging in a vertical position. At the same time, we had installed two pulley blocks whose pulleys were attached on one side to the extremities of the two shears closest to the monolith, and on the other side to the point of the needle. The main pulleys of this apparatus wound themselves around the fixed cylinder MM, and came thus into the hands of two sailors, who let them out little by little.

This first step finished, there remained a difficulty to solve. The monolith, held back by derricks whose ropes were pulling its tip at an acute angle, had a tendency to retreat back, eastward. In order to keep it from sliding in this direction at the moment that the little trunnion came out of its encasement, we built on the block itself a masonry construction in mud. The face of this wall on the side of the excavation was clad by three circular pieces whose radius equaled the distance between the roll U and a rounded cross-piece fixed to the middle of the base with iron clips.

All these preparations being finished, the rotation took place, and the trunnion could be seen rising little by little and coming out of its grooves. During the movement, a ball of sap, pressed by the weight of the mass, compressed and penetrated the brick wall. Consequently, because of this compression, the monolith began to press on the ends of the planks. Instantly, the beams bent and sank down into the sand, forming an arch pointed toward the sky. The obelisk turned repeatedly around an axis that drew closer to the pyramidion and finally reached the foot of the vertical center of gravity. In this position, it was in equilibrium on a plane inclined away from the towpath.

We lowered from the top of the shears four new pulley blocks that grasped the base of the obelisk. The main pulleys of this apparatus were linked to little capstans, each controlled by eight people.

Helped by the action of these "motors" that tended to lift the base, the obelisk inclined further. Then cracking sounds were heard in the various parts of the

The lowering of the obelisk, as drawn by Léon de Joannis, second-in-command on the *Luxor*

encasement, and soon all our efforts were unable to lower it to the point where we wanted it, that is to say, placing the ties of the encasement on the path, which had assumed a counter-slope in the part occupied by the first two struts (Plate II).

Having arrived there, the obelisk was only touching the ground at a few points, and almost ruptured because of its own weight. It was urgent to do something to remove it from the excavation and place it as soon as possible on the planks. To this effect, we removed the restraining cable that was encased in a belt around the cladding. A lever attached at its extremities was carrying mobile pulleys of four derricks, whose fixed pulleys were attached on the riverbank side Z. These main pulley blocks were wrapped around the same number of capstans, each moved by forty-eight men.

On November 16 this apparatus, as well as the four derricks that supported the base, were put into use. All eyes were fixed on the obelisk, which remained immobile under this enormous traction. The men were encouraged to push more vigorously on the bars, enlivened by a "Hurray!" At this signal, the pulleys screamed, or as the sailors said, sang. But the wire stretched too much, and soon the maneuver was interrupted by the breaking of the two apparatuses.

It was an irreparable loss in this circumstance. The monolith had to traverse a distance of four hundred meters, and from the beginning, without having gained one centimeter of ground, we would be losing part of the mechanism that was indispensable for us to take it to its destination.

The force transmitted to the monument by the simultaneous action of the system was double what should have caused it to break. The resistance that opposed the movement forward thus came from some barrier encountered by the cladding on the frame of planks on the towpath. As it was, this cladding was not accessible at any point; it was entirely buried in the sand, between two piles of debris which barely allowed a person to move around the monolith. There was no other way to reach it than to remove the shear legs, which caused the further problem of creating an escape route for the sand that was being so strongly compressed by the weight of the mass. This could cause a collapse, which would have caused serious accidents.

A piece of the shear had penetrated forty-five centimeters into the wall. It was so strongly stuck to the bricks that we had to use a carpenter's chisel to remove it. The soil (of the bricks) was chipped off, bit by bit, under the tool of the worker as if he were working knurled oak. Those were true mud shavings, whose temperature resulting from the enormous pressures they had undergone was so hot that you could not pick them up with your hand.

When the piece was removed, my surprise was enormous to find that the rings of the wood had detached from each other. They resembled a spyglass.

Although prepared in advance for problems, I was

a little disconcerted seeing the mess that had occurred both in the ground and in the scaffold. One could say that the greatest difficulties were awaiting us at the end of the lowering, and that our real stumbling block would come in the process of carrying.

The first section of flooring was tilted, the girders and the floor had broken in two parts. The fragments of the cables still attached under the girders, and lifted in the middle by the sand, came against the crossbeam of the cladding. This opposed the motion, and created an insurmountable obstacle to moving the load of the monument.

My first concern was to remove, little by little and with great caution, all the sand that was under the flooring (Plate II, figure 3). While people were doing this with a cautious slowness, since it could only be done through the opening made to extract the piece of the shear, two carpenters were busy shaping bars of the capstans in the shape of a wedge. With the help of these bars, pushed in and hammered between the cladding and the fragments of the flooring, we succeeded in lowering the planking below the cross members. Added to the force of the capstans was the action of ten jacks that were under the foot of the obelisk. These helped the efforts of the engines by the vertical thrust of their screws.

These preparations concluded, we began the process. It was that much more important because we had put everything we had into it. It had to succeed, and for that to happen, we had to act quickly, promptly, and with coordination. The more we delayed, the more difficult the task became.

Fig. 3.

Cross-section of the vertical projection after the 2nd rotation.

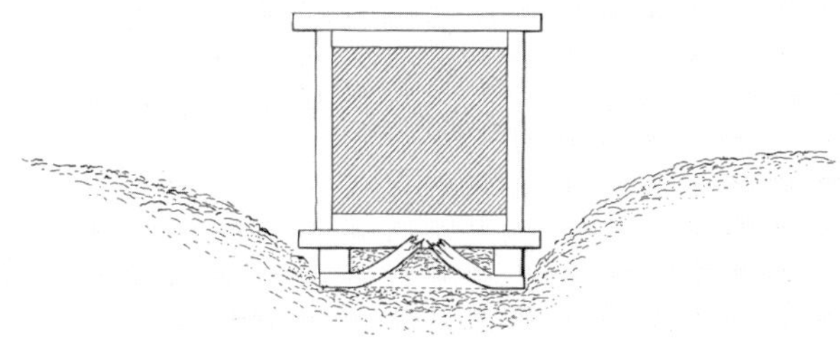

Detail from Plate II, figure 3: Sand was removed in order to stabilize the foundation while lowering the obelisk

At the command "All in motion," the apparatuses at the top and bottom, as well as the jacks, were put into play. Five minutes later, the ropes had the maximum degree of tension that they could withstand without risking breaking. I was about to stop the men when, to my great satisfaction, I saw the obelisk start moving and make a sudden one-meter lurch. All the parts of the segment of the towpath were again messed up. We had to remove the sand that had accumulated beneath the planking, dig in more wedges to lower them, and start the same work again.

At every movement in the progression, the same obstacles appeared, accompanied by some new incident. The obelisk jerked to the right because of the uneven slope of the terrain.

Before fixing this problem, it was much more urgent to get the monolith out of this difficult and dangerous place. So the operation continued, and after extraordinary efforts throughout the day, we finally managed to seat it on its planking, and remove it entirely from its excavation.

The three planks being engaged, the towing was of course stopped while we waited to get the necessary materials for the construction of the fourth section.

These were taken from the wood that resulted from the demolition of the scaffold of the shears, which was no longer needed. The two upper beams were used as girders for this planking, which was tied to the front one as described above.

We then proceeded to transport the monument by sliding it with the help of four derricks attached to the same number of capstans. The moment had come to figure out how to line it up so that its axis would coincide with that of the path from which it had deviated considerably. This was achieved by means of several props placed with their points on the faces of the cladding. These pieces of wood forced the mass to move left or right and reciprocally, according to their position, and to turn on its center of gravity while it was advancing toward the bank.

These difficulties conquered, the rest was nothing more than an affair of patience and time.

The operation would be long and hard, even more so because we had to slide the monument along a weak wooden pathway, placed on ground that settled unevenly. With each forward movement, it was necessary to reposition the main stations. Every seven meters we were forced to move sections from the rear to the front and to pack down the terrain again. Finally, we needed to reposition the capstans and the setting points of the pulley blocks several times. However, our first trials made us almost certain that we could drive the obelisk up to the ship. Therefore, we got busy with the preparatory works for its introduction into the hold of the *Luxor*.

The ship had been placed in alignment with the two obelisks. It presented its prow on the side of the monolith that we wanted to load first (Plate V). The top of the hull *LLL* that had to support it formed an extension of the plane of the towpath. In this setup, at first glance, nothing should be easier than to create an opening in the front of the ship large enough for the passage of the obelisk. But the rebuilding of this wooden path would have demanded a large workforce and a great quantity of grained wood, which we were completely lacking. We could only remove from the support platform twelve planks that were damaged by the bolt holes. These materials and the remains of the sections, that is to say, eight beams of 0.18 m. square and seven meters long, comprised all our resources in construction wood. The carpentry personnel was reduced to five able men. With these elements we had to finish this operation, so well begun, before the rising of the water.

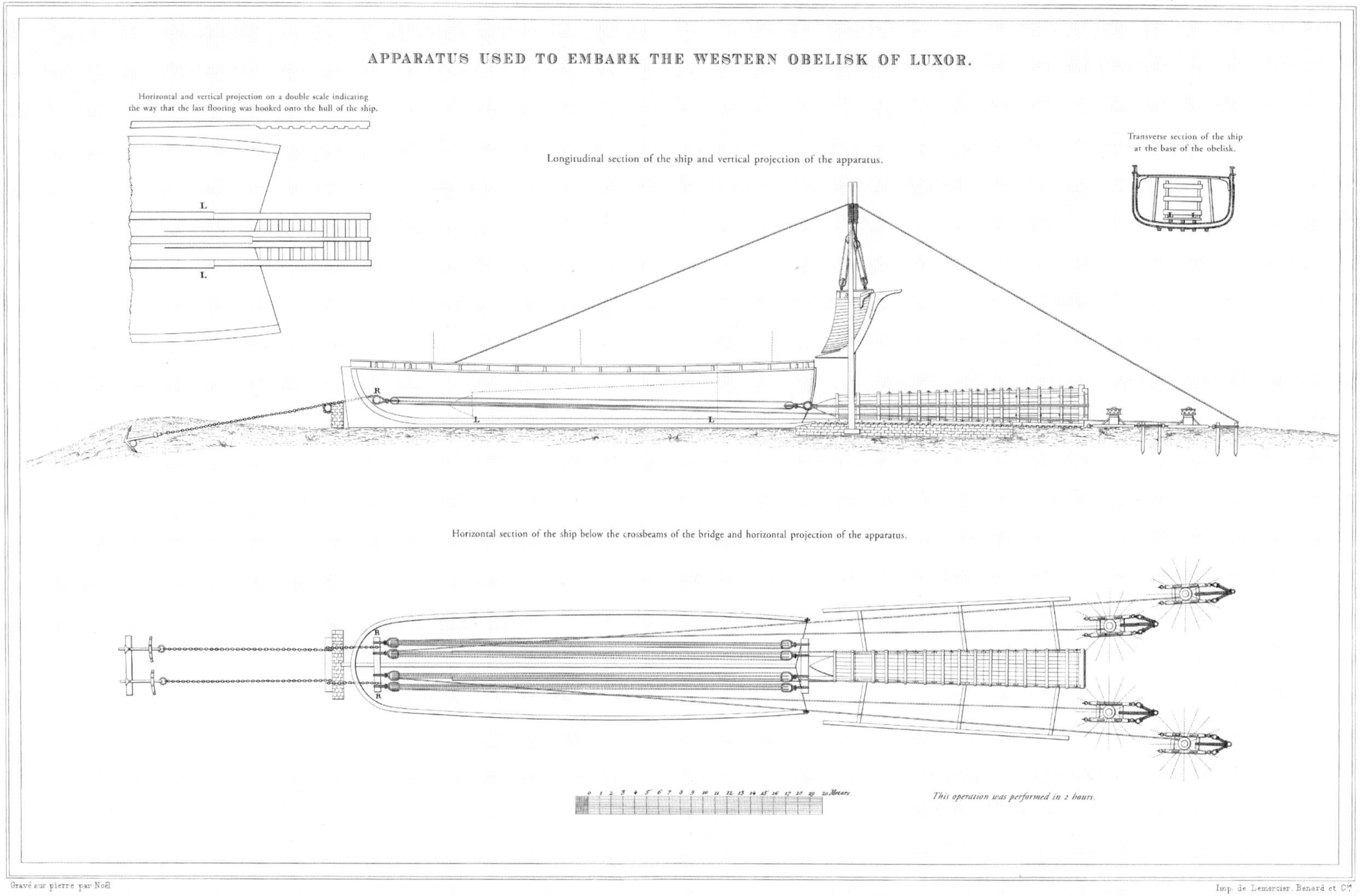

Plate V: Apparatus used to embark the western obelisk of Luxor

Instead of breaking down the front of the ship, we detached it by sawing across at the foremast. This slice, kept as a whole, was lifted out with the help of two shears put together in a St. Andrew's cross and placed to the left of the path.

While we were busy with these works, the obelisk continued its progression. After going about four hundred meters it presented its pyramidion to the cut that had been made in the ship. We then joined the front of the towpath with the two parts of the hull, *LL*, in such a way to form a continuous plane with the girders of the hold. We then proceeded to set up the apparatus for lifting the obelisk aboard the *Luxor*.

Setting of the Apparatus and Loading of the Monolith (Plate V)

Two anchors were dropped at the rear of the boat and placed symmetrically on each side of its longitudinal axis. At their anchorings, or buckles, were tied two pieces of chain. The chains passed through holes made in the hull and were attached inside on a roller, RR, which held the fixed pulleys of four derricks whose mobile pulleys were tied on a swing bar attached to the two extremities of the obelisk's rope belt. The tops of these apparatuses were attached to the same number of capstans, each moved by forty-eight men. The brick wall set against the sternpost was designed to prevent the chain from raking on the ship and to maintain the retaining system in case the anchors gave way under the force of the traction.

On December 19, 1831, with everything set up like this, we started the capstans, and in less than two hours, the obelisk was loaded aboard the *Luxor*.

It would be difficult to depict the joy, the satisfaction, of the workers and sailors. This success, so ardently desired, partly repaid our pains, and made us forget that we were nearing the season when the southerly winds would make our sojourn in Egypt unbearable.

Although he had been warned ahead of time of the day when the loading would take place, the governor of the Said supposed that, by arriving in Luxor the following day at daybreak, he would still be able to witness an operation that should have demanded several days' work. Great was his surprise when he saw that the obelisk was gone from the shore. "I did not believe," said he, "that it would have been possible to bring on board a ship such a considerable mass. The enterprise seemed to be undoable, and I can assure you that I was not alone in Egypt in holding this opinion."

On December 25, the removed section of the ship was put back into place, and the two parts fit together perfectly. The trace of the saw was barely visible and the ship seemed restored to its original state. The Turks and Arabs could not conceive of such a sudden change. They looked at the boat and at the place occupied by the obelisk again and again; they doubted what they had seen the day before. They did not trust their memory and wanted to see the reality by making sure that the obelisk was indeed aboard the ship. Finally, convinced by their own eyes, they attributed these results to the intercession of demons.

The *Luxor* did not experience any noticeable shifting as a result of the loading of the obelisk. I must attribute this important result to a path eight meters long built in front of the cut, with stones and mud from the river. This piece of the forehold, having not given under the weight of the load, perfectly protected the extremity of the ship. It was also very helpful in holding the obelisk along the axis of the ship.

After four and a half months of sleepless nights, of fears, of anxieties, the problem was resolved. To achieve the solution, it had been necessary to move more than 90,000 cubic meters of sand, take down two mounds of debris, tear down thirty houses, build several others on the roof of thc temple, and finally build all the amenities needed for the daily lives of 140 men. And all this in a country that did not have any resources, in the middle of the sands of the desert, amid the swirls of the burning dust, and under a sun that made the Réaumur thermometer reach 50 degrees.*

One single thing remained to do: to secure the slice of the boat to the rest of the ship. In order to do this, we loosened the even-numbered sheathings, inside and out, and replaced them with others whose middle corresponded to the saw mark. After this repair, the ship was almost in the same condition as when it left Toulon. However, it seemed wise, for added safety, to place inside the hold several coils of rope tying the three keelsons (a piece of wood parallel to keels to reinforce the hold) to longitudinal pieces that ran all the way to the front.

For this we used the same pieces of wood we had already used in lowering and loading the obelisk. These remnants, which because of their size and state of deterioration would have been thrown away in a normal military post, seemed insufficient for the repair. But the master carpenter, Elias, had become accustomed since our stay in Thebes to making maximum use of everything in order to supplement our lack of material resources with ingenious resourcefulness. By handling these pieces so as to create the least waste, he managed to get the number of planks that we needed.

This last step, considering the small number of workers, could only be done slowly, and by its nature it did not demand my continued presence. I had full confidence in the master carpenter, and had been reassured by the precise execution of my instructions. So, after giving detailed instructions and starting the work off, I decided, following the invitation of Sherif Bey, to visit the part of Upper Egypt situated south of Luxor. The pasha intended to create canals for this province.

Up to this moment, I had had only very little time to dedicate to scientific research and to the examination of the antiquities. I had paid attention to the objects outside my mission only briefly and sporadically. Moreover,

* Translators' note: This is the equivalent of 62.5 degrees Celsius, 144.5 degrees Fahrenheit.

this task had already been accomplished by the scholars who had accompanied Bonaparte in his brilliant expedition. The memoirs that make up the great work published from this campaign compiled the most precious documents on Egypt, from Alexandria up to the First Cataract, where the French army stopped. The upper country is less known. Reassured by the results of my work, I took the opportunity to look more carefully and attentively at the monuments that are in that province. With no claim to giving a complete description of them, I put together a few indications so the reader can have an idea of them.

Part 2
Excursion in Nubia

Chapter 1

Esna — Funerary rites — Ancient tombs of al-Ka'b — Merriment and thoughtlessness of the Arab — Temple of Edfu — Night thieves — Gebel Silsila — Kom Ombo — Aswan — The Arab Baraka — Cleverness of the Arab — Mosque of the holy man — The cataract — Philae — Temple of Girga — Sebua — Amada — Derr — Ibrahim — Whirling sandstorms — Caravans in the desert — Abu Simbel — Colossal statues of Ramesses — Maharraqa — Dakka — Dendur — Kalabsha — Qertasi — Debod — Return to Philae, to Luxor

I embarked with my servant on a *canjieh* protected by five Arabs. After casting a last glance at Luxor, we set sail on January 9, 1832. The sailors, while pushing the *canjieh* into the river, intoned their favorite song, *Habouzale, habouzelphi.* The mainsail was raised and a favorable wind took us in several hours past the island so rich in vegetation situated opposite the sanctuary of the temple. The following day at 2:00 we were admiring the upper portico of Esna,* composed of eighteen columns with flared capitals with various ornaments sculpted with leaves, stems of reeds, branches, palm leaves, bunches of fruits, leaves, buds, lotus calyxes, grapevines, etc. This portico, so richly and elegantly decorated, reminds one of pompous descriptions of the Temple of Solomon. The columns are 5.23 meters in diameter. This is all that remains from an immense temple whose precious ruins serve as storerooms for the Turks. Upon arriving, I went to the *ma'mur*† to give him a letter of introduction from Sherif Bey. Several officers of the *Luxor* had, during their voyage to Nubia, encountered some difficulties crossing the First Cataract. The *ma'mur* received me with distinction and congratulated me on the happy ending of our works. I did not know enough Arabic to hold a conversation on any other subject, and the dragoman of the *ma'mur*, although a former employee in the house of King Murat,** did not know a word of French. My servant, seeing my predicament,

* Village situated on the bank of the Nile fifteen leagues from Thebes and built on the site of the former Latopolis.

† Prefect of the province.

** Translators' note: Joachim-Napoléon Murat was king of Naples from 1808 to 1815.

reminded me that he could "babble" in Italian, and that he could, if necessary, act as interpreter. I immediately put his talent to the test. Unfortunately, he spoke the bad dialect of Genoa, almost unintelligible to the dragoman. Often hand gestures came to the rescue of the words, which greatly amused the *ma'mur*. The idea of mirth was so much associated with French in his mind that he ended all his sentences with a burst of laughter. Besides, he showered me with kindness and tokens of friendship. While giving me an order for the *nazir* of Aswan he said, "With this, people will send you up the cataract faster than the water goes down it."

After this meeting, I went toward a house from which I could hear wailing and moaning. I recognized the same cries that I had so often heard repeated in Luxor during the cholera. Wanting to observe more closely the funerary customs of the country, I examined from the doorway the people who were surrounding the body of the deceased. A woman with hair all disheveled, her breast covered with ashes, was rolling her big black eyes, beating the air with her screams. Her whole soul seemed to exhale in long, painful lamentations, and with a voice interspersed with sobs, she addressed the usual words to the deceased:

> Dear and unfortunate son, what did I do to you? Why do you leave us? Didn't you have enough to eat? Did you find the water of the Nile too warm? Didn't I take care every day to refresh it? What will I say to the girl you were to marry?

These utterances normally last from the death up to the moment of burial. A little time after, the convoy proceeded. The cadaver was carried on two palm branches by four men. It was preceded and followed by the priests, family, and friends; then came the women, each one covered by a large cotton kerchief. The priests chanted a canticle and the women responded to each verse with painful sounds. They convulsed their bodies from top to bottom, and stretched their hands toward the deceased, in the same gesture that was used to call someone. Having arrived at the mosque, the men went inside and set down the deceased. Prayers were addressed to Muhammad; meanwhile the women danced outside, passing each other back and forth. The closest relation can be identified by her more pronounced movements and more energetic gestures. She throws her arms to the sky, moves them around forcefully, and also moves her hands. Some of them remain immobile, crouching against the wall. The procession then starts again, going toward the cemetery, where the deceased is put in the sand. Then there is again the scene of the screaming and utterances that I had witnessed before they left the house.

I left Esna, the former Latopolis, where one can still see the remains of an old port and dock. I wandered along the shores of the Nile, stopping at all the temples erected on both banks. Six leagues from this town, a sailor offered to take me to the tombs dug in the rock. I started out with my guide. After an hour and a half, he took me across a vast, ancient cemetery located in a plain not far from a village that he called Heins, pronouncing strongly the aspirated H. This cemetery, roughly of rectangular shape, is surrounded by brick walls of very large dimensions. One enters through two openings oriented east and west. Numerous excavations show that the peace of the dead had been violated several times. Soon we reached a mountain in which tombs similar to those I examined in Thebes had been dug. (I found myself, without knowing it, in the caves of Eileithyiaspolis.) Most of these monuments are preceded by a room in the shape of a semicircular vault. The surrounding walls are covered with paintings, very remarkable because of the nature of the objects they represent. In the last one that I visited, there is a series of different agricultural activities. Men are cutting wheat with the scythe, quite similar to that which we use in our fields. Two others form the stalks into bundles, and they give them to a third, who ties the bundle. A fourth transports it to the circular grinding area, where the oxen trample it. Two peasants follow the animals that turn in a circle. One whips them and the other sweeps. Then come the people who winnow the wheat, in the manner used in the south of France.* The stripped grain is carried in sacks or large baskets held by two long pieces of wood. It is poured into measuring devices and displayed in heaps in front of a public official. After these scenes, you see the method for harvesting flax. There one pulls up the plant, here one forms the bundle; to the right, it is gathered. It is then carried to a machine to remove the grain. The form of this machine differs little from that of an A on its side. It is held toward the middle by a crescent. At the extremity of the upper branch is a sieve or comb into which one puts the stems of the bundles. The workman pulls the bundle of flax toward him by putting one of his feet on the horizontal branch of the instrument, and the grain falls to the ground. Nearby stands a servant carrying a large fan that he waves in order to cool the water contained in jars, which are resting on a jar stand of a very elegant shape. A little further on, some men are plowing with oxen. Others are sowing seeds, leveling the ground, digging, etc. Then come all the animals used in agriculture. Next a large quantity of fish, the cook who grills them, the fishermen who pull the net, the salter who opens the fish with a knife on an inclined bench. Geese hanging on hooks as we see in our restaurants, boats loaded with merchandise arriving on the banks of the Nile. We then proceed to the weighing. A workman writes something down on a note pad. (The weights are ring-shaped.) A load pulled on the ground slides on pieces of wood. One particular part of this last scene draws the attention. The driver, director, or engineer seems to come out from behind

* When the bundles have been crushed by the horses' feet, men throw into the air as much as they can pick up with a pitchfork. The wind carries away the chaff and the grain falls to the ground. This operation is called winnowing the wheat.

the mass of people, perhaps shocked by some mishap of the maneuver, his face contorted, both arms thrown in the air, his hair standing straight up on his head. He seems to speak to the workers, who look astonished and drop the rope they were pulling.

After having spent six hours in the desert, it was with extreme regret that, burdened by the heat and tormented by a strong thirst, I went back to the boat, where the cries and joy of the sailors indicated they had been very anxious about my absence. The Arab likes to give these kinds of shows of attachment, emotion, and gratitude, and I myself never receive them without pleasure. I deferred my departure to the following day. The sailors right away jumped to the ground, running here and there, picking up pieces of wood to make a fire. They had a pot for cooking lentils, the only food they eat on the way. The supply is quickly exhausted, which reduces them to eating *dura* bread, which is a grain similar to millet. The sailors, after washing their hands, crouched around the one large wooden dish into which they had poured the lentils, and each one helped himself, using his hand like a spoon. This meal normally lasts only seven or eight minutes. Once again, we wash our hands, and the pipe that is used by the entire crew is passed around, from one to the other. The ship's boy, during all this time, takes care of replenishing the fire around which the sailors gather. Then start the tales, the singing and dancing. The *loustic* (joker) tells very funny stories and the audience laughs. The stories of the *rais* are grave and touch on more serious topics. In general, money, the sultan, and the tarboosh (red hats) play a very important role in all the storytelling. When these subjects are exhausted, a sailor starts a song, to which all others answer, clapping their hands rhythmically. These songs greatly resemble our church songs. The bass horn is replaced by the low sound obtained from two reed pipes and from the vibration of a skin glued to the top part of a large clay funnel that they hold under the arm and hit with both hands. The Arabs became more and more animated, the songs increase, become noisier, and their joy is at its height at the moment when the storyteller, leaning on a stick and biting his shirt (the only clothing with which the Arabs are covered), appears in the middle of the circle, striking more or less indecent poses. He pirouettes on his stick and amuses his comrades with his grotesque mimicry. These scenes are repeated every day we dock.

That evening the Arabs' supper lasted a little longer. I offered them part of my provisions and, as an exception, a glass of wine each. All their faces radiated pleasure, showing through their expressions how excellent the liquor forbidden by the Prophet seemed to them. It was a pleasure to see all those jaws at work, all those hands busy, and there, as in so many other circumstances, I could see with what ease the Arab forgets, in a few moments of pleasure, the pain and fatigue of his life. In this respect, there is in him a striking analogy with our own national character, and it is for this reason that we call him the French of the Orient. The boss, however, remained faithful to the law of Muhammad. I pointed this out to a sailor who, hiding the glass in his two hands, answered: "The Prophet will not see me." I will add, to complete the topic of the practices and habits of the country, that if they violate the precepts of the Qur'an with respect to wine, they observe it rigorously with respect to prayer.

The next day we left at three o'clock in the morning. I visited the temple of Edfu, preceded by a gigantic pylon that dominates like a citadel the town built on the side of a mountain. The interior of each of these constructions contains four stories of rooms, and one goes through a spiral staircase within a rectangular enclosure whose large side has eight steps and whose smaller side has five. There are forty landings. The temple, although encumbered by sand and surrounded by modern construction, eclipses the village, mountains, and all the surroundings. It is more than 170 meters long, and this vast edifice, built with stone, is covered with sculptures outside and inside.

At a short distance from its enclosure wall there is another temple in ruins whose details are still very interesting. A series of scenes represents a woman seated, nursing her child, from his birth up to the time of weaning. In the next to last scene, he is standing, rising on his feet to take his mother's breast. Next, this same child, who has just been weaned, plays with his friends in the middle of the sands. All these phases of life are perfectly depicted.

After having hurriedly visited these ruins, so rich and imposing, I started off again, around five o'clock in the evening. At sunset, the wind fell entirely. We stopped three leagues from Edfu next to a *saqya** to pass the night. I was very surprised to see a boat under construction—if one can give this name to a shapeless carcass. It was an assemblage of planks notched together, half to half, and nailed together obliquely, serving all at once as sheathing, bottom, and ribs; the stern post was vertical and the stem post had a great protruding prow. A very crudely sculpted bird was perched on the capping. The interior only received a caulking of linen. This is, in a word, the simplest approach.

The day of the 20th offered nothing remarkable; in the evening, as we were moored on a sand bar, the screams of *Harami* (thieves) were heard around two o'clock in the morning. My servant, a young man of seventeen years of age, and nicely built, fearful of falling into the hands of Bedouins and of undergoing a fate he would rightly have considered degrading,† was knocking vigorously on the door of my room. I got up quickly. I ran onto the deck armed with my gun and I learned that we were threatened by thieves. The *rais* assured me

* A machine with cups to draw the water from the river.

† This young man had already been the object of special attentions on the part of the *ma'mur* of Esna. I mention this fact only because it leads to an observation, unfortunately all too accurate, about an oriental custom, namely that, among the Turks, it is often the means by which many young men reach high position.

that he heard them and told me to shoot in order to scare them. His fear seemed to me imaginary, and upon my refusal, he threw himself on the ground and asked me insistently to fire off some shots—and many *(kitir)* of them. I agreed to his pleas and the noise reassured him. To justify his fear, he told me that naked thieves, their arms and legs oiled, daggers attached to their feet and all their joints, would often go onto the boats in the night and take away everything they could find. They are excellent swimmers; if you shoot at them, they dive suddenly and remain, if necessary, under the water for a very long time, in such a way that it is impossible to tell which way they went. This way of stealing has been confirmed by several people, among others by Mr. Linant,* who was surprised by a gang of those bold thieves.

The double chain of Libyan and Arabian mountains which form, to the right and left, the embankment of the Nile, gets closer and closer to the river as one goes up its course and arrives on its banks at the quarry of Gebel Silsila, where we landed on January 21. These quarries are open-air. Vast excavations can be seen, whose walls are cut vertically, sometimes stepped. In some places the mountain is dug to form underground caverns. The entrances inside are decorated with porticoes, columns, and entablatures, executed with great care and covered with hieroglyphs. Even the layers of which the stones would be composed have been marked, if they had not been removed en masse. The quarrying seemed to me well planned and done with art and method. One finds traces of soundings at various points. A few defective parts of the rock where they had started to work seemed to have been abandoned before they finished, probably because they would not have been able to get blocks as large as those usually used by the Egyptians. Everything shows the care and skill of the research that they used in the choice of layers to exploit. The quantity of materials extracted is immense. These enormous cavities once held in their bosom the temples, palaces, and columns that cover all of Lower Egypt. The position of the quarry is fortuitous. Two mountains made of very strong sandstone, perfectly resistant, join at the same place on the banks of the Nile and seem to have been placed there next to the means of transportation, to inspire in a hard-working people the taste for the grandiose that they carried to sublimity.

Between Gebel Silsila and Aswan (Syene) one finds the temple of Kom Ombo, dedicated to the crocodile. It is a vast edifice, part of which was destroyed by the river, which undermined the foundations. The façade of the grand portico is perfectly preserved, and presents a remarkable particularity in its construction. It is formed of five columns, in such a way that the vertical axis coincides with that of the central column. This circumstance forced the architect to put in two

* Engineer of bridges and roads of Upper Egypt. This title is perhaps surprising in a country that is completely devoid of roads and bridges, but this at least shows the goodwill of Muhammad Ali.

doors, without a transom, instead of a single one, as one sees in all the other edifices of this kind. The colors of the cornice are still vibrant. The architecture of this monument, its columns, its flared capital, its grand proportions, its magnificence, offer a rather theatrical picture. One should observe that this temple is, compared to most Egyptian monuments, of modern construction. It was raised during the time of the Ptolemies, which might explain the singularity of its portico. Anyway, there, like elsewhere, the layout of the modern edifices seems better conceived than the old ones. But this comparison for art is far from holding true for hieroglyphs, and as to the proportions of materials, it is clear that this has already started to degrade.

Up to here, the bottom of the Nile is silt, but approaching Aswan, one sees a few black masses that surge out of the middle of the river. These granite rocks, polished by the current, become more numerous and larger as one approaches the town. On our arrival, two *canjieh*s flying the Turkish colors greeted us with rifle shots, to which we responded. I recognized the boat of Mr. Linant and that of Mr. Botta, physician to His Highness. These gentlemen, whom I had seen in Cairo, heaped kindnesses upon me. A dinner served Arab style was immediately set on Mr. Linant's *canjieh*. Although this way of eating is not very agreeable for a European, I was so bored with my way of life, with the eggs, rice, and never-ending dates prepared in a hundred different ways by my servant, that I found this meal delicious. What made it priceless for me was to find myself, after such a long voyage, among some of my compatriots. Beyond Thebes, 1,200 leagues from France, we were happy to retrace its past, which was enhanced by the happy noise of a few bottles of champagne. Among the guests was one of the chiefs of the Arabs of the desert called Baraka (providence) who was supposed to accompany Mr. Linant in his search for gold mines that were said to exist toward the shores of the Red Sea. This man, tall, robust, with a beautiful black man's face, an expression of spirituality in his physiognomy, the top of his head covered by a white turban of the same color as his robe, was examining me very closely. Mr. Linant, teasing him about his observation, told him, "This little man is the engineer on the obelisk." Surprise appeared on his face. He got up, came to sit next to me, touched my hand, asked questions, and listened with the keenest interest to the details of the process of taking down and loading this monolith.

"What an incredible operation! You should take a trip in the desert. I would accompany you there and I assure you that you would not lack for anything." I thanked him for his kind offers and, pursuing the conversation, I told him that one of our men, who had gone hunting, had not returned and that we didn't know what had happened to him. "In the desert," said he, "we would have easily found him. Footprints in the sand are characteristic signs for us that we never miss." Then he told me several anecdotes, among others the following one:

"Someone had stolen a parade saber from a Turk. The feet of the thief had been imprinted on the sand just as he left his house. This sign was sufficient for us to know that he did not belong to the oasis. Here and there we covered these prints with clay pots in order to keep them from dust that would have erased them. As we were carefully following them, one of the most experienced observers declared that the thief was carrying the saber in his left hand in the beginning and that later he moved it to his right hand. At some distance from there he announced, "Now he doesn't have his saber any more." There was a well right there. We searched it and indeed found the crucial evidence. Still using the same method, we arrived at his oasis where he lived, and the inhabitants, upon seeing the footprints, showed us the individual, who was arrested." I expressed to Baraka my astonishment and my doubts about the truth of this story. "This surprises you!" he said to me. "However, you surprised us even more with your great operations." Mr. Linant confirmed with another story the sagacity of the Arabs in recognizing men by their footprints. A young Arab who was accompanying him in the desert had asked permission to go visit his mother. "She just passed by here," he said. "I see it from her footprints." He left and came back three days later saying that it was, indeed, his mother. He found her at her sister's place, who lived in the oasis.

Whatever these stories, it is well known that the Arabs, always working barefoot and on sand that is often soft and uniform like snow, can easily recognize an individual or person from the tracks or prints that he leaves behind. It is such a widespread truth that most of the thieves, when pursued by the police, walk backwards, jumping left and right, and by all means try to erase them. This observation helped us later to return to the spot in the desert that we had searched for in vain.

Baraka is a man naturally clever, with grandeur and independence. He is frank, without flattery toward the Turks who had caused him many problems and worries. Upon returning from the Red Sea, he intended to go see the pasha to express several complaints about the authorities of the country. If Egypt had possessed a great number of such men, one could not doubt that its political regeneration would be easy and near.

The day following my arrival I visited the granite quarry at Aswan and an old cemetery where every grave is marked by a vertical stone covered with inscriptions. This quarry, located on the western side one league away from the river, offers a picturesque aspect. It is a vast desert where blocks of granite with bizarre shapes appear here and there, sometimes regular, flat, or slightly cylindrical. The excavations are filled by the sands carried by the *khamasin* (desert winds). As I said earlier, speaking of the quarry of Gebel Silsila, the proximity of the water and the soft slope which leads to it would surely have helped to transport the enormous blocks that were used for the construction of the temples, of colossal statues, and of the obelisks. Nature seems, by the way, to have accumulated these materials in this position under conditions most favorable for their exploitation.

The fine pink color of the granite of which they are made, dotted with spots of black and white, gives warm and varied tones to the reflected light and gives the sculptor the means to create on the faces more lifelike and real features. I will not talk about the method used by the Egyptians for the exploitation of the granite. Their methods, of which there exist numerous traces, have been described in many works. The cuts made in the bedrock, intended for wedges used to detach the blocks, are as fresh as if they had been done the day before. Their color had not changed at all, and they are particularly remarkable for the sharpness of the edges. The rock from which a colossal monolith weighing eight hundred tons was taken is covered with chisel marks that are oblique and parallel. They occupy a surface greater than seventy square meters. The bottom of those cuts still shows a white line as if the point of the tool had just struck it.

There is at Aswan an obelisk whose axis is inclined thirty degrees to the horizon and is connected to the mountain only by the lower plane. The worker removed from the rock only the portion of granite strictly necessary to work the vertical sides. The trenches that run around the monument are only eighty centimeters wide. It seems that these monoliths, like the Paris one, were not always exempt from defects, and because of that, they were careful, as one sees on the Aswan obelisk, to probe on all the sides where there were hairline cracks, in order to see if the cracks met or penetrated inside the stone. These probes were done with a tool similar to the drill used by modern sculptors.

Several blocks bear hieroglyphic inscriptions that probably refer to the use or volume of the mass removed. These signs are not engraved deeply. They are merely stippled with a pointed tool. This observation led me to examine more carefully the ones that are sculpted on the obelisks and on the walls of the sanctuary. I noticed that everything that was in relief in a hollow was worked in the same manner. The curved faces show on their top a series of small circumferences, whose rough surfaces were smoothed by polishing the granite. Working the hieroglyphs demanded, on the part of the artist, a patience that would horrify our French vivacity, but which should hardly surprise us when one considers that the Egyptian people transformed huge rocks into palaces, would devote a whole chain of mountains to tombs for its royalty, and would provide, as dedication for its edifices, obelisks that themselves are huge and imperishable monuments. Our chisels would dull very quickly on this granite because the workman would remove, with every stroke, pieces that were too large. If he took the precaution to strike more lightly, as it was done in front of me, the chisel would last for half a day. Thus, without saying that the Egyptian tools had an incomparably better tempering than ours, one can easily explain how they worked granite.* They would first mark with the chisel a shallow trace of the contour to carve. Next they would remove the interiors with a needle without touching the outside edges, which must be kept very sharp, the chips never encroaching on those first lines. If this is not exactly the procedure followed by the Egyptians, I am sure, because of the experiments I did, that this process would lead to the same result. The granite is easily polished with grit or pumice. It is in this way that, before my eyes, two small obelisks were carved that I brought back from Luxor. The polish is in no way inferior to that of the ancient obelisks, and the edges are no less sharp.

To the right of the quarry, in the direction of the river, there is a hill on which is the mosque of a holy man. While heading that way, I met on the road a man wearing a green turban, suffering from asthma that only allowed him to breathe with the greatest difficulty. This was the holy man himself, who took me to his dwelling. In front of the door was a circular platform, perfectly swept, whose circumference was encircled by rough stones. It is forbidden to the infidels to cross this barrier, which rises forty centimeters above the ground. The Arabs themselves can only go inside the circle after having removed their slippers. This mysterious place is a kind of Arab confessional. The penitent lies down on his back, both hands above his head. If he rolls like a cylinder, it is proof that he did not commit any sins during the day. In the opposite case, he broke a few rules of the Qur'an. My pilot, Barmarin, rolled perfectly, which didn't seem to impress him, because shortly afterward he told me, "It's a trick. You can roll however you want." The holy man, still out of breath, barely able to articulate a few sounds, kept a grave but comical bearing. I said my goodbyes to this holy person, to whom I gave a few piasters, which smoothed his frown.

I had taken the precaution of asking the *ma'mur* of Esna for a letter to the *nazir* of Wadi Halfa, residing at Aswan, in order to avoid delays in the passage of the cataract. This functionary had the pilot summoned and gave him the order to take me immediately to Philae. This Arab, of great stature and wearing a beautiful red turban on a rather nice head, which was unfortunately ruined by a chin covered in scabs, embarked on my *canjieh* and drove us with much skill through the rocks, in the middle of which the Nile snakes, fast and turbulent. These rocks, fairly high and fairly sharp, cross its bed for a league and a half. This passage is dangerous and difficult. The twists are very numerous; the path of the navigation changes with the height of the river. According to the season, one must sometimes be towed with a rope; sometimes one must lift the boat out. The crossing lasts about two hours. The appearance of the cataracts and their surroundings is truly frightening.

* Translators' note: Lebas is wrong here. The ancient Egyptians only had bronze chisels, which were too soft to quarry granite. They used dolerite, a stone even harder than granite, to pound the granite from the quarry.

One navigates around the enormous blocks of granite that the waters have polished like a mirror. The banks are lined with yet greater masses, which pile one upon the other, rising in the shape of a pyramid to a very great height. A few sparse palm trees here and there contrast with this harsh and wild landscape. A tortuous current, a few waterfalls, rocks bizarrely cut out in all directions, irregular points that one needs to avoid, entirely grab the attention. One would think oneself transported into the bony framework of the earth, far from all kinds of living beings, when, suddenly, one discovers pylons, colonnades, perfectly regular lines, finally a succession of striking monuments, sublime products of the highest civilization. This is the island of Philae, which separates Nubia from Egypt.

To explain this multitude of rocks that encumber the river, the Arabs say that a pharaoh threw them from the mountain in order to stop Nubian boats that were coming to attack Egypt. The pilot of the cataract, who is usually clad in rags that make him look like the old Charon, had worn his good clothes in my honor, for which I had to pay a lot. As a result, which usually happens in this country, although I was supposedly piloted for free, I ended up paying more than the travelers who were not referred by the authorities.

At Philae, as at Kom Ombo, the modern temples are better laid out than those of the previous centuries, but the hieroglyphs are sculpted with much less perfection. The flat stones that cover the portico of the monument situated to the left of the great temple are joined obliquely—in other words, the joins next to the angle are askew, something that I did not notice in any other place.

I devoted two entire days to visiting this island, which is nothing more than one of the rocks of the cataract. It has today, for inhabitants, only two or three families of fishermen. There, as I already mentioned, Upper Egypt ends. The country further upstream takes the name of Nubia. There also ends the Arabic language. It is at Philae that the French army stopped. An inscription that we could not contemplate without vivid emotion recalls this glorious page of our history. It is worded thus: *The French came all the way to here. Here the division of Desaix, under the orders of General Bonaparte, pursuing the Mamluks, forced them to take refuge in the desert.*

I left Philae on the 25th. The wind was favorable. I passed the palace of Kalabsha and slept in Kirchen, where there is a temple dug into a mountain of sandstone.

The portico is supported by four columns and eight pilasters, against which are huge statues that represent Ramesses standing, the head covered with the symbolic *pshent*, his arms crossed, holding in one hand the flail of justice and in the other the scepter of command. The monumental pose of these statues, their rigid immobility, impresses on the whole edifice a character of greatness and seriousness. It is from there, perhaps, that the Greeks got their idea of the caryatids.

To the right and to the left, niches of rectangular shape are dug into the wall, each containing three standing figures. The first smokes a pipe and the last is a young woman. The façade of the mountain in which the entrance door is cut is at an angle, in effect forming a torus whose convexity is turned outside. Going inside one finds a great room whose ceiling is supported by six pilasters much higher than those of the portico, against which are also colossi similar to those outside, and perfectly preserved. Below the navel is a head of Isis with cow's ears, as one sees at Philae and Dendara. Little children, pressed one against the other, cover her nudity. On the lateral walls are eight niches containing the same figures as those outside. At the end, facing the entrance, is a second door leading to a room of the same width as the preceding one (13.7 meters). It is completely unremarkable. There are two square pilasters and four niches. In the middle is a door which leads to the sanctuary, at the center of which is a stone altar, and in the back a niche where four persons are seated.

On the 27th we arrived at Sebua. The temple seen there, whose construction is attributed to the great Sesostris, is almost buried in the sand. An alley of sphinxes precedes the pylon. Up front one sees two men in the position of soldiers learning to march in step. One crosses the pylon and enters a room in ruins, supported by ten pilasters with statues of Ramesses resting against them in the same position as we saw previously at Kirchen. This room leads to another that seems dug into the mountain, and its encumbering sand makes it impossible to explore.

From Sebua to Amada is about ten leagues. The Nile makes a large elbow in this space and goes northward. As the winds are north to south, one is forced to drag the boat with a rope. These sailors, without asking me, replaced the tricolor flag by that of the owner of the *canjieh*, the *nazir* of a factory at Qena. To impress the peasants even more, they dressed properly, put on red tarbooshes, and went to recruit in the name of their master some peasants, whom they brought along, hitting them with palms. It seems that the villages located on the right bank, amid a rich plain covered with the most beautiful vegetation, are exempt from certain laws on the condition that they haul the boats of these Turkish officials going upriver. These poor people, hitched to a rope, were tormented and mistreated by the eight sailors, who took joy in imposing, in turn, their own show of power, of which they are so often themselves victims.

The temple of Amada, situated on a rise, is buried in the sands. One enters through a covered gallery at the end of which is a portico supported by twelve pilasters and four columns. These supports are arranged in four rows. The last row is formed by the columns rising in front of the back wall, where the door of the temple is cut. The pilasters have very small sides and are finely finished. The interior is composed of a first rectangular room leading to three other deeper rooms. The one in the center continues all the way to the back. The other two are cut by a transverse wall forming four separate chambers.

This temple seems to belong to one of the most remote epochs. The cartouches carved on its walls belong to Thutmose III. Thus, the work of the hieroglyphs is more perfect than in any other building. They are generally of very small size and done with exquisite care and taste, which places them in the *belle époque* of Egyptian arts. They rather look as if they have been cast in bronze and engraved rather than sculpted.

From Amada I went to Derr. This village is situated on the right bank of the Nile, almost in front of Amada. There is in that place an oven that belonged to the *nazir*. It is the only place up to the Second Cataract where one can bake bread. One obtains this authorization by giving a present to a functionary, who at the same time sells you the wheat.

I witnessed in that town the purchase of several black slaves whom ships brought from the interior of Africa. They were taken to Cairo, but the merchant would stop anywhere when people wanted to buy some. Here, as several interested buyers had come, they presented to them the black women, who were locked into a common room in the back of the ship. The buyers examined them with the greatest care, looking at all the joints, the head, the eyes, the breasts, the neck, the nails, the ears, the feet. They walked in front of them. They were turned and turned again, the way a horse dealer haggles over horses at the fair. During this examination the face of the buyer shows perfect indifference, the total absence of all feeling of humanity. Everything for him seems to come down to a simple deal, where his only preoccupation is not to be cheated on the material value of the object bought.

At a short distance from the dwellings, there is a temple dug into the mountain. It is preceded by a portico and the inside plan is almost the same as that of Amada. Only the first room is longer and supported by six pilasters. The three rooms on the left upon entering are not on the same axis: a setting that was determined by some defects of the rock that covers it. At the rear of the last room, in the middle, one can see fragments of statues.

Going up the Nile, one discovers a mountain that rises perpendicularly at the edge of the river. Walls of dry stone indicate a rather large village. A few ruins of Arab architecture were the only ruins that I found on this summit. My walk had been long and difficult, but I was amply rewarded by the imposing view that the surroundings offered. There are two deserts separated by the Nile that offer, from different viewpoints, the terrible reality of two completely sterile natures. The desert on the right is dotted with pyramidal hills between which roll torrents of dust swept by the *khamasin*. They remind one of the sight of the surging sea spilling on the rocks; these sands are deposited and accumulate on the sharpest point of the rocks, on almost vertical planes. Sometimes they form themselves into a staircase, slightly concave in width and whose perfectly sharp edges shine like a luminous line.

In the left desert there are several moving mounds that disappear at a certain distance and give way to an immense beach of sparkling sand. One would not find a single pebble in its entirety. From this desert arise at intervals enormous spiral columns that turn on themselves with incredible speed. The instantaneous formation of these swirls, the speed of forward motion at which their material particles move in two perpendicular directions, the vaporous cloud that results from their rupture and follows the direction of the wind, the luminous rays reflected or refracted by the diaphanous masses, offer one of the most magnificent spectacles to hit the eyes.*

I don't know what fear gripped me upon seeing the children of the desert go with a prodigious speed toward the path followed by the caravans. The imprints of the feet of the camels that unfurled in long ribbons on this sea of fire reminded me of the tragic end of a small caravan that an eyewitness had told me about during my stay at Syene. I could almost see in action the oriental tale of this Arab. I was following anxiously the comings and goings of these unfortunate people, looking for the traces of those who had preceded them. One gust of wind changed everything. The mounds changed in aspect and shape. The ground undulated like the water of the Nile at the cataract. Three times the travelers neared the edges of the river, from which they were separated by only a few leagues, and three times they reentered the desert thinking they were on the wrong path. Because of these comings and goings, their provisions were used up. The last drop of water had just come out of the water skin. These unfortunates, dying of hunger and thirst, at risk of being engulfed in a forest of sand columns, made the ill-fated decision to sacrifice the animal that might still have saved them. The camel, the bird of the desert, was cruelly slaughtered. The Arabs avidly brought their inflamed lips to the pocket of the stomach, to draw the remains of the liquid that this quadruped is careful to store. Soon they fell to the ground, and their cadavers, dried up by the ardent burning fires of the sun *(nar kitir)*, after a few days resembled the mummies of Thebes. "For

---

* Here, as the traveler Bruce recounts in his memoirs, are the effects produced by these whirlwinds of sand:

"At one o'clock," says he, "we reached a bunch of acacias where we stopped. From there we had the most magnificent spectacle, as well as the most terrible, that one can see. From the depths of the immensity surrounding us there suddenly arose, as if by enchantment, a great number of sand columns; sometimes they were carried with extreme speed, then remained immobile. At other times they would progress slowly and with majesty. They pounced on us as if they were going to crush us; already grains of a hot sand were hitting us in the face. Then they would recede up to the horizon. Their tip would then grow and get lost in the sky. Sometimes it would even separate itself from its base, and the column would not exist any more. Other times they would dissolve with an explosion. At the crack of dawn they got closer to us. They marched in five rows and at a distance of three miles. The strongest of them seemed to have a diameter of twenty feet. The wind, which suddenly turned to the northwest, chased them away from us. Any escape would have been in vain. The most agile horse could not have avoided the death with which they threatened us. This is why we watched their terrible movement in silence and immobility."

me," continued the narrator, "I was too attached to my camel to imitate my companions. Full of confidence in its natural instinct, I left him entirely free to move about. Several times he raised his head and glanced at all the points of the horizon as if he were trying to find the direction that he should follow. Then he started off without hesitation, and after three days of heavy walking, he took me to Aswan."

Thanks to the intelligence and philanthropic zeal of the pasha, humanity no longer has to lament such unhappy events. The compass is now used in all of Egypt. Every caravan chief who has this instrument knows its properties and easily orients himself, even in the most critical circumstances.

Leaving Ibrahim, I traveled for two days without discovering anything remarkable. I had familiarized myself with the conversation of the Arabs, their songs, their dances, the profile of the banks of the river, and the customs of the natives of the region from my stay in Egypt and they did not offer anything interesting for me. Boredom began to overtake me, but I was approaching Abu Simbel. The view of two temples carved into the rock soon reawakened my attention.

What could be the purpose of these two monuments, built into the heart of a mountain that rises steeply on the banks of the river, in a country absolutely barren, where one does not come across the slightest trace of vegetation, old or new? Mountains, struck by an eternal barrenness, extend quite a distance. The intervals that separate them are, like their surfaces, full of sands brought by the desert winds, which twice blocked the entrance to the great temple. Thinking these thoughts, I turned to the mountain that led to the façade. When I saw it, everything changed. My attention was totally captivated by the imposing spectacle that appeared to my view. I fell into an ecstasy of admiration in contemplating the four colossi seated to the right and left of the entrance. They represent Ramesses. No difference can be seen between them. They are as identical as the rock from which they had been sculpted. But such is their beauty that one cannot stop admiring them. Their features are of such perfection that nothing equals them. The expression of them is, so to speak, living. Seated, his brow calm, the great Sesostris appears to be resting in his glory and enjoying his own satisfaction. His soul itself is in all of his features. It seems to say to those who would look at it, "I have done great and important things. The people who surround and importune me are happy for me. I am happy through them." The attitude of the body is in harmony with these sentiments. Several statues are sculpted between the legs, and several at the side, notably that of a standing woman; all her hair, turned to the side, falls in the form of a parabola on her shoulder. My servant, a simple young man without any knowledge of art, and with a mediocre intelligence, was not insensitive to this spectacle. "Oh!" he said to me. "That one looks like a good man! How happy he is, sweetness and kindness are painted on his face!" Such is the power of art when it is brought to its highest expression. It seduces, it captivates even the man who is only moved by the simple instinct of nature. I doubt that the chisel has ever produced a tableau of such great interest and with such a prodigious effect.

Main dimensions of this monolith	
Total height of seated colossus	19 meters
Length from one shoulder to the other	6.9 meters
Length of foot	4 meters
Width of the nose under the nostrils	7 meters

The statue of the woman of whom I have just spoken, although her size is 3.5 meters, placed as she is near the colossus, resembles a small child.

The façade of the temple, carved into the mountain, is framed by a torus molding; above is the cornice on which are dog-headed apes standing in the same position as those that decorate the bases of the obelisks of Luxor (see Plate IV). Over the door there is a large standing figure with the head of an owl carrying an ankh in one hand. At his feet are other statues of medium height. One enters a large room supported by eight pillars, against which are statues of Ramesses standing, with crossed hands, holding the scepter in one hand and the flail of justice in the other. The prenomen of Ramesses is sculpted in hieroglyphs below his navel, and the vine branches of our statues are replaced here by serpents whose heads are crowned with the image of the sun. These figures are as perfect as the colossi outside. Their height is 6.7 meters and the top of the crown is 1.5 meters from the roof. All the walls of the room are covered with very remarkable scenes; I will give as an example the fight between two kings. Here, as in all the temples, the sovereigns are represented in colossal dimensions. The size of the conquered king is a little less than the conqueror. The latter, standing in his chariot, holds him with one hand, as with the other he pierces him with a spear, whose handle would hide a part of his face in an ordinary position, but the artist wanted to show the complete profile of the conqueror so the weapon passes behind the head. Could it be in memory of some great victory achieved on this very place by Sesostris that this temple was dedicated? That would explain the choice of such a deserted and isolated site.

The room was 15.6 meters wide and 17.35 long; to the right on the lateral wall are two doors which lead to two chambers four meters in width by fifteen meters length. At the back of the room there are three doors. The middle one leads to a rectangular room whose ceiling is supported by four pilasters. It is 10.8 meters in width and 9.2 in length. The other two lead to two corridors; each of them is pierced on the right wall by an opening through which one penetrates into two very long rooms whose axes are not perpendicular to that of the edifice. All around there is a bench, above which are excavated niches which were probably intended to receive the goods and offerings that were deposited on the bench. In the middle of the floor is a band of hieroglyphs. The center room, of which we have spoken, is followed by another one not as deep. One crosses it to reach the two lateral chambers and the sanctuary located in the axis of the temple, where there is a monolithic altar and four statues in the back. The interior of this edifice receives daylight only through the entrance doors. As a result, with the exception of the first room, all the rest is in complete darkness.

The second temple is smaller. The sloping façade is divided in five parts by six buttresses. In the space that separates them are sculpted figures of the standing gods. The front of each buttress is covered in hieroglyphs. The interior is made up of a first room with pilasters whose capitals are decorated with the head of Isis with cow's ears and the hairdo of Egyptian women. The second room leads to two lateral chambers, and at the back the door of the sanctuary bears the statue of a god.

I was only five leagues from the Second Cataract, but I did not think I needed to go up there, having heard through specific reports that it did not offer anything remarkable. Anyway, I was pressed for time. I was awaiting instructions from the government that might have required my presence in Luxor. Also, going up the river, in order to take advantage of the wind I had passed several temples that I intended to visit on my way back.

I stopped three days at Abu Simbel. After a last look of admiration at the colossi and the façade of the Great Temple, I started back to Luxor. I arrived on February 4 at Maharraqa. There remains from this great temple only a kind of portico, open to the sky, supported by eight columns. One reaches the floor of the gallery by a circular staircase, perfectly crafted. This one work would suffice to prove that the Egyptians knew descriptive geometry. It is one of the drawings that one does at the École Polytechnique.

From Maharraqa I went to Dakka, five leagues away. The palace of the same name is in ruins. It is preceded, like the other edifices of this kind, by a large pylon. In the portion that is still standing, the hieroglyphs are perfectly preserved. Vultures, their wings outstretched, shining with the most vivid colors, decorate the ceiling. In the battles or triumphs, this bird accompanies, leads, or precedes the heroes. One penetrates into this part by a door without a lintel placed between two columns. Then one finds two other rooms, the last of which communicates with the outside. A scarab, the symbol of the soul, replaces on the cornice the winged globe, which usually occupies this place, and it produces a very beautiful effect. A broken-down enclosure wall surrounds this edifice.

We were progressing toward Dendur when the sailors stopped rowing to show me the most beautiful, most lively, most agile, and most graceful animal. It was a herd of gazelles coming to drink in the Nile. They were playing and frolicking on the sand like herds in the meadow. The sailors suggested that I go on the land if I wanted the pleasure of shooting them. I accepted their suggestion and they went and stood at a distance. I was going to walk toward the gazelles, to examine this pretty animal more closely, not to hurt it. I was still two gunshots away when, because of a slight sound, they left with the speed of lightning, and in spite of the efforts and shouts of the Arabs, they disappeared in less than a minute. On returning to the *canjieh* I found the *rais* busy heating a nail, with which he burned himself on purpose on both thumbs to cure his scabies, with which both hands were covered. This remedy seemed a bit extraordinary, but what seemed even more extraordinary was the patience with which he bore the pain. This resignation is one of the traits of the Arabs. During our stay in Luxor, someone brought to the doctor of the ship a fifteen-year-old boy whose calf had been torn by dogs. The surgeon cut, chiseled the flesh, and made numerous incisions, and the patient did not scream once during the entire operation.

I arrived at Dendur. This temple, located on the left bank, is preceded by a terrace surrounded by a curved wall imitating the shape of a torus whose concavity is turned toward the river. At the opposite extremity rises a triumphal door, similar to those of Karnak. Next comes the façade of the temple. The first room leads to a corridor, which, flanked by four chambers, leads to the outside of the edifice and arrives at a square hole dug in the mountain. It would be extremely difficult to give an explanation of this arrangement, which appears very mysterious.

Since my departure from Abu Simbel, I had not visited any village. At Kalabsha I found, as in all of Egypt, a great number of naked children, waiting for me on the banks, crying "baksheesh" (alms). A mother was arranging the hair of her daughter, putting into it a paste composed of soil of three different colors and some fat. All the Nubians use this pomade to preserve their hair from the strong sun. This strange cosmetic does nothing to improve their looks. The sweat dissolves some of the soil, which runs along their temples, stays there, and adds quite a bit to their natural ugliness.

The Temple of Kalabsha is near dwellings. A part of a dock built in stone on the bank of the Nile serves as a revetment to a flat dirt surface with a terrace, which is bounded on the right and on the left by two lateral walls. This terrace is in front of the temple; it communicates with the square by a path, which is really nothing more than the continuation of the terrace, reduced in width by two indentations in which stairs have been made to go up on the road. The dock on the side of the river has the same shape as the one at Dendur. During the flood, one disembarks on the terrace. When the waters are low, one disembarks on the ground and walks to the staircase to arrive at the temple.

Such a grand avenue built with stones of such great dimensions indicates that Kalabsha was a first-class temple, one of the most frequented in Egypt. Indeed, because of its size and the number of buildings of which it is composed, it can be compared to Karnak.

Three exterior enclosure walls set at large distances from one another surrounded it in all directions. The doors have been cut out. One finds in one of these walls several chambers of the same dimensions. The main body of the edifice is in ruins. After having crossed the pylon, one enters a vast courtyard. To the right and left, there is a gallery in which only one column remains. This gallery stops at a certain distance from the lateral walls and reveals a new façade formed of four well-preserved columns. It leads to a room also supported by four columns. One then enters several rooms much smaller than the previous one. The parts still standing are covered with hieroglyphs, sculpted with great care, and whose colors, after so many centuries, shine with a brilliance that still excites our admiration. All these ruins are witness to the immensity of the edifice. The materials that are still there would be adequate to build several palaces and justify the idea that we have formed of Egyptian genius and works. These people, indeed, seem to have taken upon them, in the execution of their monuments, to produce nothing but the grandiose. They equated moral greatness with material greatness, and if one can say that the first can sometimes do without the second, it must still be admitted that nothing could touch and strike the soul more deeply than the intimate union of moral and material greatness. This observation is more easily appreciated if we think back to the time when these monuments, which for us today are merely objects of art, commanded the respect and worship of the population.

Going back up the mountain, one finds a little temple dug into the rock that is composed of only one room and a sanctuary. Three seated statues occupy two niches chiseled in the back wall. A similar niche also exists in the sanctuary. The avenue of this temple is excavated in the rock. The two sides of rock that are left exposed are decorated with a tableau representing combat sculpted with a perfection, a precision in the details, that one does not find anywhere else. Chariots, horses, fighters, everything seems alive and in action.

At the moment when I was going to leave Kalabsha, I realized the *rais* of the boat was being held against his will by several fellahs who were talking agitatedly and refusing to let him go. "He bought milk," they were saying, "he has to pay." My servant, still thinking that he was dealing with the Arabs in Luxor, threatened to hit them and he was arrested. Feeling the necessity to intervene, I fired my gun in the air to frighten those fellahs, but it was in vain. I had to negotiate with them and I only managed to free the captives by paying what they expected from the *rais* and more, several piasters in baksheesh. I recognized in this circumstance, as I have observed several times, that the Nubians are in general less servile than the Egyptians. They maintain a certain spirit of independence and are not treated by the Turks with such harshness. It even seems that they actually receive the portion of the harvest to which they are entitled. Thus, they are better dressed, less worried, and less fearful in front of their oppressors. Their lands are better cultivated. One sees in them a certain volition. This altercation ended, the oars put in place, we headed to Taffa, where we spent the night. In this place there is only a portico with six columns, worked with as much care and perfection as those of Esna. There are almost no sculptures. From Taffa I went to Qertasi, whose temple is almost completely in ruins. One can see from a certain distance the remains of a portico; the capitals of the columns are decorated with vine and lotus leaves. The central axis is about ten meters. The entablature is formed by a single stone of the same length. I arrived in Debod the same day.

The temple is preceded by three triumphal doors, similar to those of Karnak. The façade is formed of four columns and two sloping back walls bordered by a torus. One enters by a door without a lintel situated in the middle in a rectangular room. On the left lateral wall, there is a door that leads to a chamber jutting out of the façade. This room is unique; there are no others like it. The wall facing the entrance is pierced by three openings. The middle one leads to three successive rooms, the last one being a sanctuary where a monolithic temple in granite rises.

These rooms, condemned to perpetual darkness, these nooks carved in the wall attached to the sanctuary, separated from the entrance door by several partitions, everything says that this edifice was consecrated by mysteries whose veil could only be lifted by Champollion the Younger.

This was the last temple that I had not visited going up the Nile.

Back in Philae, we stayed there a few hours waiting for the pilot, who eventually appeared, all dressed up. After a thousand salutations and compliments, he told me, as an introduction to the second passage that we were going to make through the cataract, that the previous day a large English *canjieh* had sunk there. Nobody had drowned, and the Arabs saved from the boat all that it carried. Having reached a narrow channel in the rock where the water was seething with force, he started to cry and to prance like a man who believed himself in danger. I realized that the purpose of these antics was to scare me in order to get a higher salary. I was not duped. I quietly took my revenge by telling him that these little undulations were not anything compared to those of the sea, which rise sometimes as high as the nearby mountains, and come to break against the sides of the ship with a crash and a frightening noise. This comparison put an end to his fake demonstrations. Shifting his thoughts totally somewhere else, during the whole trip, he only occupied himself with insinuations toward my servant, who came to tell me about it. I made it clear to the pilot that if he continued his misplaced advances, I would have him lashed with two hundred blows of the whip upon my arrival in Aswan when we disembarked two hours later. I gathered some provisions and, following the course of the river, we arrived three days later in sight of the village of Luxor.

With the lower masts removed, surrounded by a double row of mats and covered in sand, the ship *Luxor* presented the aspect of a ruin in the middle of ancient ruins. A foreign observer, having no data on the rising of the Nile, could never have imagined that this mass of wood, 150 meters from the river and stranded on a beach five meters above the current level of the water, carrying in its bosom a colossal monument, would soon be lifted and transported to France across the Mediterranean and the ocean. Although prepared in advance for the periodic regeneration of the Nile, one could not help feeling a sudden fear in view of the large sheet of fluid that this phenomenon would necessitate.

Chapter 2

On the formation of the land of Lower Egypt — The Nile in ancient times did not water the Valley of Egypt — Opinion of Herodotus on the subject — Geographical and political divisions of Egypt — Taxes levied with the whip — Army conscription — Tendency of the Arabs to steal — Nile floods — Fertility of the land — Happiness and skill of the Arabs — Their sobriety — Muhammad Ali — His preoccupation with war keeps him from organizing the country — Habits, character, and clothing of Arab women — Marriage ceremonies — Almees — Dance of the Bee — Maghribis and snake charmers — Courtesans — Unchanging climate and atmospheric phenomena in Egypt — Its influence on man and animals — The *khamasin* — Imposing grandeur of the monuments — Voyage to Rosetta — New meeting with Muhammad Ali — Rising of the Nile and launching of the *Luxor* — Its departure for Alexandria; return to France — Arrival at Toulon

After having gone through all of Egypt and part of Nubia, it seems to me that the time is right to say something about the topography of this country. It is confined between two chains of tertiary sandstone mountains whose direction is generally north and south. One, on the east, borders Arabia; the other, on the west, Libya. This narrow and tortuous valley is cut perpendicularly along its length by two shoals of granite which divides it into two distinct basins—that is, the lower basin, which constitutes what is properly called Egypt and ends at Philae; and the upper basin, or Lower Nubia, which spreads till the cataract of Wadi Halfa. Both basins are irrigated by the Nile.

In its upper course, this large river is formed of the confluence of two rivers called the Blue Nile and the White Nile. The first comes from the country of the Agows, and takes its source, according to the traveler Bruce, in a marshy plain located more than two miles above sea level. The second one crosses the unknown regions of Africa. After their junction near Halfay,* the Nile flows straight to the northwest over the space of two degrees of the meridian. There it receives a tributary on the right, the Tocazze,† which comes from Abyssinia. It then travels without receiving any additional water, either from rivers or from rainwater, through Upper Nubia. From there it descends into Lower Nubia through the cataract of Wadi Halfa, breaking for the last time

* Translators' note: Modern Khartoum.

† Translators' note: The Atbara River.

across the rocks that rise from its bed at Philae. Then it reaches the beautiful country of Egypt and flows into the sea, after a course of more than six hundred leagues.

Today, as in the time of the pharaohs, the Nile fertilizes all the land that it bathes in its muddy waters. Everywhere else, Egypt has only an arid and lonely soil, naked mountains whose uniformity is broken by nothing living, immense sandy beaches where the eye loses itself and thought saddens.

The principle of the fecundity of the Nile is due to the silt that it carries. This silt is deposited partially on the inundated land, partially in the bed of the river; the rest rushes into the sea. The first deposits are visible, easy to notice. The land that has just been inundated is covered with a layer of black soil whose thickness increases with every inundation. This is a phenomenon that no one can doubt. The land in Egypt has thus necessarily undergone a change in elevation. As for the deposits that take place in the water, it is easy to prove that they are the obvious consequence of the first deposits. Indeed, the height of the inundation has been the same for a thousand years. The land is irrigated today as in the past; thus the bed of the river has risen in proportion to its banks. One can argue as to the thickness of these layers of silt, but the fact in itself is indisputable.

Experience confirms the correctness of this observation. The bases of a great number of monuments, as much in Lower Egypt as in Upper Egypt, are several meters below the level of the river. At whatever depth one has dug, one always finds material identical to the alluvials of the Nile. This second result, like the first, proves that the land and the bed of the river have risen at the same time and in the same proportion. To give but two examples of the thousands that prove the fact, I will cite the two colossi of Memnon erected on the plain of Gurna. The bases of these statues have almost completely disappeared under the layer of silt. They are five meters below the actual land, which is, in fact, inundated every year.

We have seen that the beaching bed of the *Luxor* was level with the upper plane of the pedestal. The inundation went 2.7 meters above this point, which itself is raised 2.6 meters above the courtyard of the edifice, in such a way that if the palace were cleaned of the modern debris that surrounds it, it would be totally covered today by the water up to the height of 5.3 meters.

The first result proves that a difference of level of at least five meters has been formed in this part of the valley. The second result proves that the bed of the river has risen nearly the same amount.

Although it is impossible to assign a precise date for the erection of these monuments, if one thinks about the centuries that must have passed before the Egyptians had even thought about raising their prodigies of architecture, one would have to give those alluvial deposits such a considerable thickness that it will seem exaggerated to any judicious man. And thus, one will inevitably be led to conclude that Lower Egypt has not always been watered by the Nile, that this river has not constantly deposited silt. In order for us to measure the variations of the waters, we will examine what is happening today at the Aswan cataract.

At the strongest part of the inundation, the Nile entirely covers the rocks that form the cataract. One month later, one sees, surging from the bosom of the water, the tops of the highest granite peaks, whose number increases in proportion with the fall in the water level. Then the waters come and pound them violently and polish them through abrasion while flowing through the spaces separating them. These intervals diminish more and more as the level approaches their bases, where they form a continuous chain that is situated around four meters below the low-water level. The height of the rising being nine meters, a depression of the land thirteen meters in depth in the bed upstream from the cataract would establish the upper level of the river against this dike, and Egypt, where it does not rain, would be dry. This question would be resolved if one could make sure that at that depth the components of the soil are the same as at the surface of the bed. But this proof is given by the thickness of the deposits (five meters) which happened at Thebes, in a period of time certainly less than one-third the period between the [Biblical] flood and now.

Thus, one cannot deny it; the Nile did not always flow into the Valley of Egypt; there was a time when, stopped in its course by a transverse dike, it flowed back in its entirety into Nubia and bathed in its fertilizing waters the deserts of Libya. Then, these immense sandy beaches cut by numerous canals were covered with rich vegetation, inhabited by an industrious, rich, and happy people, which, from all the evidence, was better governed than today. This state of affairs must have lasted up to the moment when the successive deposits of the river in its upper bed raised its level above the continuous chain that supports the rocks of the cataract. Then, a part of the water flowed violently through the spaces that separated those rocks. This incessant action removed a few pieces from the rocks, and created small openings as it escaped. Soon these grooves became more pronounced canals through which the water flowed with a speed proportional to the slopes. It is only from this time on that the Nile started depositing silt in the lower basin by gushing into it from a great height. This height must have diminished gradually in proportion with these embankments, whose mass is increased by each inundation. However, it was still very considerable in the time of the pharaohs, since all of antiquity agrees and talks of the last cataract as a prodigious falls whose terrible noise rendered the inhabitants deaf.

As a greater volume of water flowed through Egypt, the height of the rise diminished in Nubia. By consequence of this lowering, the highest plateaus ceased to be inundated, whereas the low lands of Egypt were watered and rose progressively. Thus, there was in the fertility of the two countries a succession of variations caused by those of the river, which brought about, little by little, fertility and civilization in the lower country. Abandoning the plateaus that became uncultivable, because of the water receding, human beings went down to Egypt where they could obtain, without effort, unparalleled harvests from the virgin land. They followed the course of the Nile, which had conquered the spacious valley that gave birth to Thebes and its monuments. From an invariable nature emerged a magnificent country cultivated by a people who increased little by little, whereas the upper basin was getting poorer. It is to this era, when part of Libya was still flourishing, that the wars that broke out between the inhabitants of these two countries should be dated. Thus, these numerous armies of which tradition speaks were able to go from Egypt to Nubia and vice versa. Today, it would be impossible for troops to live or move through the middle of these vast deserts.

Such are the consequences of our inspections, consequences applicable to the upper basins up to the end of the rains, since the same causes have produced changes analogous to the flow of the waters. Therefore, we must conclude:

1. That humans have followed from basin to basin the alluvial deposits of the Nile, which took them first to Nubia, where the river was stopped in its course by a transversal dike;

2. That, at the time, Egypt, deprived of fresh water, was in part, if not totally, covered by the sea and completely sterile;

3. That the Nile, having opened a passage through the cataract, deposited in the valley a considerable amount of silt, which throughout the centuries came from the bosom of the water, and has thus formed fertile lands, which the Nubians seized.

Such is the origin of Egypt and of the people who left behind such a long heritage. This origin is confirmed by the tales told of the priests of Sais to Herodotus, and then, through the observations of this historian,* by the testimony of a great number of ancient writers who thought that Egyptian institutions came from Ethiopia, and by the paintings that represent men of this ancient race. We will end this geological overview with some observations about the soil of the Thebaid, which we have examined with particular attention.

The elevation of the land under discussion is greater on the banks than in the plain. A longer stay of the water on the edges must necessarily increase proportionally the quantity of silt deposited. This mass of alluvials, of which the land of the Thebaid is composed,

* The priests of Sais told Herodotus that in the time of Moeris, the first pharaoh, all of Egypt, with the exception of the nome of the Thebaid, was nothing but a marsh, and that there was no visible land below Lake Moeris, although it takes seven days to sail from the sea to the lake. Any judicious man, adds this historian, will see that it is the same for the whole country located above the lake, to a distance of three days' sailing.

Under the reign of King Moeris, a rise of eight cubits was enough to water Memphis, and nine hundred years later, at least fifteen cubits were necessary. (Larcher translation of Herodotus.)

is crossed in various directions by veins of sand; they form many canals through which the water is filtered across the fields, dissolving the salts that they contain. A part of the liquid is absorbed by the silt and the rest moves according to the laws of fluids. One can see from this that several sandy veins can converge to a single point very far from the Nile and form, in a space lower in elevation than the level of the incoming water, these reservoirs that one finds in the desert and on the Quseir road. The following facts will prove this opinion. There exists behind Luxor Temple an excavation that seems to have been dug by the hand of man. One might even suppose that it was a former pond that received the water of the river from a canal that was open to the sky, of which no trace exists any more. Today the water penetrates through infiltration and follows the variations of the Nile. It is the same with the one at Karnak Temple, situated two miles from the river. The water that it contains is always brackish, but to a lesser degree than the wells at Quseir, a consequence of the phenomena that we have accepted, because the quantity of dissolved salt must, under equal circumstances, be proportional to the distance traveled by the water, before manifesting itself above ground. It is due to these infiltrations that the blocks of granite, which are not reached by the inundation, are nevertheless broken down by the mechanical action of the natron that the water contains. This salt is found in abundance in all the land of the Thebaid up to a depth of fifty centimeters.

The Arab farmers distinguish perfectly the sand veins from the silt of the river. At the time of low water, they sow watermelons on the edge of the bank that is uncovered land. Walking one day between the edges of the river and the sternpost of the ship, I saw a peasant who took possession of the land to plant it. I noticed that after having poked holes in the land meter by meter, he deliberately failed to plant seeds in several of the shallower holes. I asked him for an explanation. He replied: "Those are bad. If I were to dig a little more, the water would spurt up immediately. You see, I fell upon a vein of sand that lets the water go through." I will not discuss any more the properties of the Egyptian soil. They do not necessarily belong to my focus of attention and I must leave to minds more versed in geology to draw more complete and interesting conclusions from exact observations of the land.

Back among the Arabs, among whom I had been living for a long time, I continued to study their form of government, to observe the variation and the regularity of their seasons, the fecundity of the soil, the customs of the natives, the food they eat, their social interactions, their work, their entertainments, etc.

Egypt possesses seven governments: four in Lower Egypt, one in Middle Egypt, and two in Upper Egypt.* Each of these governments, entrusted to a Turk who

* They are also given the name Sa'id (Upper Egypt), Bahari (Lower Egypt), and Wustani (Middle Egypt).

takes the title "Excellency," is subdivided into several *ma'muriya*s, the *ma'muriya*s divided into *nazira*s, and the *nazira*s into main villages, which are divided into subordinate villages.

This vast territory of Egypt, the buildings that cover it, the Nile that crosses it, the inhabitants who populate it, all belong by proprietary right to the pasha. All of it is at his full and entire disposal; its commerce is exclusively in his hands. The concept of individual property does not exist in this country. Each governor works in the name of His Highness, the supreme authority, and transmits his orders through couriers mounted on dromedaries to various Turkish authorities. Only the chiefs of the villages are Arabs responsible for the execution of the orders; they are legally responsible.

Each fellah must pay a poll tax, a tax proportional to the area of the land occupied by his house. He is subject to the agricultural corvée and forced to cultivate and sow a certain number of feddans of land. For this last work, he is allotted a proportional part of the harvest. On the day of accounting, the agent in charge proves to him, through irrefutable arguments, that he has nothing to claim personally; that on the contrary he is in debt to the government for the distractions that he caused, and in addition he did not pay all the charges for the *miri* (tax). The poor man goes back empty-handed to his village, happy to have escaped the inevitable kurbash (beating). Such is the fate of the unfortunate descendants of the Egyptians, that the tradition of the blows of the stick by which the pharaohs paid for the gigantic works of their ancestors seems to have been more faithfully preserved than the sphinx, the obelisks, and the pyramids.

If this Arab farmer has any dealings with Europeans, either as a worker or as a domestic, the Turkish agents find out how much he could earn each day, and from that moment, they hound him, hunt him like a wild beast. It is in vain that the Arab buries his money in tombs or entrusts it to the belly of the earth; the inflexible *ma'mur* will manage, under a thousand pretexts, to extort it from him. They first start demanding that he pay the back taxes of his parents; if this does not make up the entirety of the sum that they ask of him, he gets ransomed by his friends. If he refuses to satisfy this, after three summonses, made a few days apart, the envoys of the *ma'mur* lay him down, his belly against the earth. They raise his legs vertically and keep the soles of his feet horizontal by means of a stick attached to a chain against his heels. Two individuals hold the ends of the stick and a third sits on his neck, as on a horse, to keep him from moving. In this position, he is asked again to pay. If he persists in refusing, the executioners, armed with a kurbash [whip], apply more and more blows to the soles of his feet. Soon the blood flows, and the torturers only answer the screams of the victim with: "*Hat filus*, give money." The scene gets more and more animated, the cries and pains of the victim only excite the fury of the executioners. The blows come with more speed, until the moment when the Arab, unable to resist the violence and torment any more, heaves a big sigh and throws to the ground in a convulsive motion the gold coins that he had hidden in his mouth.*

This behavior of the fellah who brings his money and lets himself be beaten before giving it presents an obvious contradiction. However, it is explained by rather simple considerations. Eager to keep his gold, the peasant is proud to endure torture and to convince, by his absolute resolve, that he possesses nothing. On the other hand, he fears not being able to endure to the end the pain to which he exposes himself. In this way he keeps the option to buy back the rest of his life if the sufferings become intolerable. He is by nature patient, and a few succeed in tiring their executioners. This known resignation is often the cause of some unfortunate people, betrayed by false reports, having to undergo the torture though they don't own a single para.

Here is a fact that I can certify, of which I have proof; it happened during our stay in Luxor.

It is the custom for the government to leave the harvest of dates to the peasants, in return for a tax. Because of the drafting of men in Upper Egypt, a grove of palm trees, which did not bear a single date that year, was without a worker. The *ma'mur*, told by his agents that an Arab scribe employed by the French was receiving a daily salary, went to work right away to take it from him. He called him, gave him his hand to kiss, showered him with compliments, praised with exaggeration his zeal and activity, and told him that for his service he would be given the harvest of the grove of date palms. The scribe hesitated and finally mumbled "No." At this word, the *ma'mur* became angry, reproached him bitterly for refusing a gift from the chief of the province, and screamed in anger: "Bring the kurbash." The Arab turned white, brought to his lips the hands of his master, thanked him for his kindness, and laid at his feet the price demanded.

The Turks themselves cannot entirely escape the rapacity of the governors. For eighteen months the soldiers who lived in Luxor and functioned as police had not been paid. In order to live, they had sold some of their clothes and weapons. Finally, pushed by necessity, they were trying to extort from the poor peasants the produce that they had been able to take from their tyrants. These are the kinds of abuses that the Arabs have to bear, imposed upon them in turn by the soldiers, the chiefs of the villages, the police officers, the *nazir*s, the *ma'mur*s, the governors, and all the servants attached to their service.

Men are recruited for the army with the same harshness. There is no random selection in Egypt. Each village must furnish a certain number of men. The officials *(cawas)* go into the houses indicated by the chief of the village, grab the conscripts, and to keep hold of them, put on some kind of handcuff that binds

* Diodorus Siculus said: "The Egyptians believed they lost out if they paid before being beaten to force them."

the two thumbs. They are then brought into the most populated village.

When their number is sufficient to fill a boat, they are sent to Cairo. This departure, as one can well imagine, creates a scene of sorrow. The mothers, the sisters, the relatives, assemble around them, scream, pull out their hair, and, weeping, say their last farewells. Unmoved by this spectacle that they don't even seem to notice, the Turks meanwhile fill the boat with their recruits, and set off for Lower Egypt.

No population has more repugnance for military service, and uses more tricks to avoid it. Often it is only with incredible difficulty that the agent in charge of gathering recruits manages to collect the required number of soldiers.

When some unusual project is contemplated, mainly clearing drainage canals, the authorities proceed by assembling a press gang within a radius of eight to ten leagues. Men, women, children, all are brought to the site and, willy-nilly, they have to work, most of the time with neither instruments nor tools. They have no choice but to dig the soil with their hands. The guards in charge of watching them keep them from leaving. The chief of every village is responsible for bringing them food, but as he almost never has sufficient means, the relatives of the workers—the few who are left in the houses—supply it. Those who have nothing die of hunger while working. I have seen one of these unfortunates working on a canal near Luxor, who for eight days had eaten nothing but reeds. He resembled a ghost. Next to him were a pregnant woman and three small children, four to five years old, exhausted from work and hunger.

Can we, after this, be surprised that the Arab is generally wary, suspicious, a thief, and careless about the future, that he tries to drape himself in misery in order to escape the painful scenes that we just described? To say that he owns money is to ensure a series of misfortunes and tortures, which only end when he has given his oppressor his last para. The vices that one attributes to the Arabs are not natural to them. They are the consequence of the inept and brutal treatment to which they are submitted. For example, would they try to cheat the authorities of their rights if they had not incessantly had to fear theft by their underlings? Would they disobey the orders of the high authority, if these same agents did not constantly beat them with a stick?

What happened to me during the first day of our work verifies this opinion. Someone had stolen the *masse* (a kind of hammer) and two shears that our stonecutter had left on the base of the obelisk. The *nazir*, told of this theft, promised to have them found. Four days later, not having found them, I told him that I was going to tell the governor of the Said via a courier on camel. The *nazir* begged me to wait until the next day. He then used all his resources. All the inhabitants occupying the houses built around the obelisk were assembled and submitted to the inevitable kurbash. Told of what was going on, I went there and asked forcefully that this cruel treatment be stopped. The *nazir* pointed out to me that the punishment had only been inflicted on the most unfortunate, and that it was necessary to inflict it on one of the principal inhabitants, leaving it to me to have it stopped. On the second blow, I begged for forgiveness and the individuals were sent to prison. They were barely incarcerated when an Arab came running, out of breath, and put down the *masse* and the shears, which he said he had found outside the village. Since then the Arabs, well convinced of the efficacy of the protection given us by the pasha, did not steal the slightest piece of equipment spread out over all the riverbank. They even brought us the nail heads.

There is no civil registration in the country. The age of an individual is unknown. If one asks them about this, they give some remarkable event to which they attach a part of their life. The removal of the obelisk, for example, will serve as a date for a great number of them.

The climate of Upper Egypt is not as variable as in Europe, except for the days when the *khamasin* (south wind) comes. The sky is always pure and calm. One never sees the slightest little cloud; the air is completely dry. It is perhaps due to this absence of atmospheric vapor that Egypt owes the conservation of its monuments. There are some whose colors, with the exception of red, which is extremely adherent, come off with the slightest touch, yet they are completely unchanged. In such a climate, one is never in doubt what the weather will be. One can count on the continuity of the work from sunrise to sunset and calculate precisely the entire time needed for the completion of the work. However, we happened to witness a momentary change in the order that Nature has imperturbably fixed in Egypt. An abundant rain fell during a part of the night. A few hours more and the habitations of Luxor would have disappeared, at least those built with mud; they would have melted and liquefied.

The Nile takes six months to rise and recede. The rise begins in Thebes on the first days of June. First the water loses its transparency, little by little. One then notices slight oscillations. A few days later, it takes on a greenish color, and then the rise becomes obvious. Later the color turns dark red, the speed of the current increases, and the water contains masses of moss. The Nile, in the Thebaid, rises eight to nine meters. The rising motion occurs without agitation, without moving any of the soil. These phenomena, which happen only on this river, cause it to be given the name Wise. Before the waters have reached their maximum, the irrigation canals are opened to relieve and spread the inundation. This is a memorable time, called in this country the "opening of the chalice," and which for the Egyptians is always a signal for rejoicing and holidays.*

* In the time of the pharaohs, it was the time of the flowering of the lotus; this plant represented the inundation for the Egyptians, and its flower, which resembles a chalice, was emblematic of the entrance of the Nile into the canals.

Generally, not all the land is covered by the Nile. The portion that remains dry gets wet by infiltration. Since the height of the river above the ground depends on a multitude of circumstances, it cannot be predicted exactly. The height is around twenty to thirty centimeters on the plain of Thebes. The periodic flooding of the river invariably determines the time of agricultural labor, which is not very painful as the soil does not require extensive preparation as it does in Europe. All it takes is sowing and covering the seeds. The first part of the ground, which emerges at the end of November, is planted with beans. The second part, left dry in December, is devoted to wheat, barley, chickpeas, and lentils. In February people eat fresh beans, and in May, chickpeas. The harvesting of the wheat and barley is done in May and June; the lentils and chickpeas in the month of April. This is also the period of flowering of the date trees. Well-developed male flowers are suspended from the branches of each female palm tree; the former fertilizes the pistils of the latter. The *dura*, principal food of the Arabs, and corn are planted in July and August on lands that have not yet been inundated, and harvested in October and November. The growing and harvesting of opium take place in March and April. The bank of the river, uncovered in April and May, is considered as free land because of the short time that separates the time of dryness from the next inundation. It is thus left to the fellaheen, who plant watermelons, cucumbers, onions, and chicory as they wish. This kind of agriculture demands very few workers. The fellah, standing, makes a hole in the ground using a stick, and drops in the grain, which he covers with his foot. This harvest is finished in two months, which is the only time the land remains dry. Almost always the farmers are obliged to get into the water to harvest the crops.

In Egypt, they also cultivate indigo, sugar cane, rice, cotton, flax, etc.

Also, the produce is not the same each year. It depends upon the speculation and the decision of the government, which orders larger or smaller proportions of the land dedicated to such and such produce, according to the profits that it hopes to gain.

The soil of Egypt is immensely fertile. The source of its fecundity consists in the silt deposited by the Nile. This vegetal soil, black as long as it remains humid, becomes, when it is exposed to the sun for a certain time, identical with the sand. I had the opportunity to make this observation several times. On the sides of the mountains, exposed slopes presented to the untrained viewer successive layers of silt and sand. A few days later the color was uniform and showed no layers; it would have been impossible for me to distinguish them. Apparently, the vegetal particles are detached from the sand by the heat and are reduced to dust that the slightest wind blows into space.

With respect to fertility, I will tell what happened to the expedition itself. I had had an enclosure wall built around a small piece of land dedicated to growing vegetables. We planted grains from Europe, which astonished us by the speed of their growth. Nine days later we had radishes that we needed to consume promptly because thirteen days later they were flowering. It was pretty much the same for cauliflowers and onions, whose tails had to be pinched in order to keep them from sprouting. A patch of beans, of tomatoes, of eggplants, is sufficient to feed the staff and local authorities of the country. We kept picking them and more always grew back. Finally, a branch of a fig tree the thickness of a thumb planted at the door of the garden became in one year a tree whose trunk diameter was ten to twelve centimeters.

According to the details that we have given on the harsh, brutal, ferocity of the Turkish government, it is astonishing to find still in the Egyptian people gaiety, spirit, incredible talent, and great aptitude for the arts. The results obtained by myself, and of which I have already spoken, confirm the truth of this assertion and are witness to the native qualities of the Arabs. I have lived with them in close proximity; I have seen them and studied them during the four seasons, which contain pretty much all the circumstances of the life of a man, with regard to the various periods of his life. What resources, what abundance of means they find in themselves! Certainly, a great number of years would not be necessary to bring them to a high level of civilization. All the arts would make immense progress in this beautiful country if agriculture were encouraged and fostered by a wise and enlightened administration. The land of Egypt, incredibly fertile, distributed in properties to the inhabitants, would produce a hundred times more, and Muhammad Ali would increase his revenues. Although he is the only true owner, he receives annually one-fifth of the production. One part is wasted by the people and the remaining is stolen by the subaltern authorities, who impose, as I have just said, all kinds of abuse on the poor fellaheen. The Turks, who benefit from having the country worked by a group of slaves, object that these unfortunates, because they have few needs, would be happy to cultivate just a few acres of *dura* necessary for their existence. This is an assertion against which we raise our voices with all the strength that a long and serious observation can give, and a deep conviction: like other people, the Arabs passionately love money, which procures the joys of life and objects of luxury. They look for all the means to acquire some. I have seen them travel fifteen to twenty leagues each day under the burning sky of Nubia for a small salary that they were going to use to buy a tarboosh. In other circumstances, they would run for hours across the plain in the hopes of finding a few tobacco plants. For a cup of coffee, they would have crossed a desert. Isn't it obvious from this that these men would work ardently at something that would give them the means to satisfy their needs, which would increase with their well-being? The viceroy knows perfectly the situation of his kingdom, the lack of ability and character of the Turkish employees who exert the power under his domination. But, preoccupied as he is in building his political power and then having it accepted abroad, it must be admitted that it is difficult for him to participate actively in the internal organization of the country. To those who pointed out to him its deficiencies or vices, he often answered: “What do you want me to do? I am alone. Later we will see.” His Highness was considering at this time the conquest of Syria, or his own country’s independence, and this moment was not favorable to upset the internal administration. Such a revolution would have infallibly attracted the hatred of all the men in power and would have added to the difficulties that he already had to overcome. Let us hope that Muhammad Ali, whose genius is ahead of his century, whose ambition is to be a figure in history next to Peter the Great and even to surpass him, will not delay and will soon emancipate a people who need only laws and a social organization to place them at the level of the most advanced people of Europe with respect to industry, agriculture, and the arts.

I will end the telling of this excursion with a few details on the customs and habits of the Arabs: If they lack color, at least they will have the character of truth, so necessary to appreciate a little-known people.

The clothing of the men is but simple pantaloons and a shirt of cotton. The least miserable cover their head with a turban or with a red tarboosh. The others wear a kind of white skullcap, which they call a *taki*.

The women who appear in public also wear pantaloons and a shirt similar to those of the men, and on the head a large kerchief that goes down the back, all the way to the bottom of the leg, and whose corners are held in the teeth to hide the face. The girls and boys, up to the time of puberty, are entirely naked. The sex is well formed, the shoulders wide, the face very regular and very expressive, the teeth very white, the eyes sparkly, of a beautiful black, protected by long lashes that radiate around like brushes. The breasts are more separated and more pronounced than in European women. The women of Egypt paint their lips dark blue, and often the part that separates them (the lips) from the chin; this symbolizes their sex. Their gait is proud, agile, elegant, and it is impossible to carry with more grace a burden on the head or a little child straddling the shoulders. I paint here the women of Thebes; the Nubians are far from sharing the same physical advantages. They are ugly, dirty, their facial features are very pronounced, and the paste that they use to pomade their hair gives off a very disagreeable smell. The bodies of the Egyptians offer nothing obvious to distinguish them from Europeans. We find bodies of all sizes, more or less well shaped. Their eyes are like women’s eyes, very sparkly, very animated, and of a beautiful black. The color of the face is also tan.

Flirting is not a sentiment foreign to the Arab women. They go out constantly covered. The young ones, when a European approaches, although seeming to hide their faces with more care, in this fictive motion manage to show themselves. They also like chatting. Forced to remain at home, all day long, they go in the evening to get water from the Nile with pots called *ballas*, which they bring home. These pots are too heavy for a woman to be able to place by herself on the head. When it is time, they gather on the banks of the river in two or three groups and, under the pretext of helping each other, they spend a very long time chatting, which seems to be a very great pleasure for them. If the Nile has risen during the day, they sing religious chants for a few minutes. When it is time to leave, they wait for everyone, take up their burdens, slowly walk back to their dwellings, and keep up their cackling. In this way, the young women are admirable. Their beautiful proportions, their graceful and elegant bearing, the light clothing that barely veils their shapes full of purity and freshness, recall with a delicious reality the beauties that imagination and art realize most perfectly.

As to the habits and character particular to Egyptian women, as they believe they have been created solely to serve the pleasures of men and to be agreeable to them, their most ardent desire is to become the spouse of a man, Turk or Arab, who is rich enough to be able to keep them enclosed indoors. The women who have this fortunate privilege form a separate category. The master to whom they submit themselves takes all imaginable measures to make it impossible for them to cheat on him. They only go out very rarely, and are completely veiled by two or three cloaks pierced by two holes that barely allow them to see ahead. But there, like everywhere else, there are locks and gates. They still manage to get free at least once a year—the Day of the Dead. They insist, as a religious principle, that they should go alone to the cemetery to pay their respects to their deceased relatives and friends. That day they have complete freedom. and there are few, they say, who do not abuse it.

The marriage ceremonies present several rather curious practices that I believe should be recounted. First of all, among the Arabs, the big question of the dowry receives a totally different solution than among us. Instead of bargaining for his fiancé, the groom-to-be gives the parents of the girl a value in money, goods, and animals, and promises in front of the qadi, in case of repudiation, to pay her in the form of compensation a predetermined amount. When these conditions have been accepted by both parties, people celebrate for several days the coming of the marriage. A sort of stove attached to a pole, filled with burning wood, is set up in front of the door of the groom's house as a symbol of light and a sign of rejoicing. They have all kinds of amusements. Someone shoots a gun. The *almee*s perform several dances that delight the guests. The mother of the bride, in the presence of a few family members, tears the virginal membrane of her daughter. The cloth used for this operation is carried triumphantly in the streets of the village and placed like a banner in the window of the bride-to-be. The young man is undressed and washed with warm water and soap. These preparations concluded, the two spouses are officially united. These practices show only the rude and barbaric, but nevertheless respectable, expression (and it is for that reason that we have mentioned them) of the importance that the Arab attaches to the purity of customs. They do not prevent, however, the slipping in these moral observances of certain deceptions to which the mother sometimes agrees.

I spoke of the *almee*s. We know that this name is given to professional dancers, a kind of nomadic dancing girl, always accompanied by one or two men and several old women who play a violin with three strings and a tambourine. They do this to earn money, and thus perform in private houses. Their dance, their performances, their movements are extremely lascivious, artfully varied, and present with great skill all that the most unrestrained passion can imagine. The *almee* holds in her fingers two small copper cymbals the size of a six-pound coin. She sings and dances, sometimes the whole night long. The quantity of brandy that she drinks then is unbelievable, and she would not be able to tolerate it if her constant movement did not encourage perspiration.

Of all the kinds of dances that I saw performed by the *almee*, I will limit myself to recalling a few characteristics of the one they call the Dance of the Bee.

The *almee* appears dressed in a pantaloon with a skirt on top, a spincer [a tunic cut at the waist] and a belt. The head is covered by a large veil that falls to the bottom of the leg. The music begins; the artist starts to dance and sing while clapping her cymbals. This is a kind of overture that has nothing to do with the main scene. The first movements are languid. The face of the *almee* is motionless. She walks rather than dances. This preamble lasts about a quarter of an hour. Soon she pretends to hear the buzzing of an insect flying in the room. She listens closely, looks, pretends to chase it, continues her steps, slips, and then declares through a painful scream that she is being chased by a bee. Then the two instrument players and her companions come near while screaming at the top of their lungs "AK-AKO!" There it is, there it is! Her movements become much more vibrant; the *almee* doubles her complaints, brings her hands to her body, sometimes on one side, then the other, turns and turns in all directions. Her face lightens up, lightning sparkles in her black and rolling eyes. She is suffering from violent pains. She squats, wraps herself in her veil, and rolls on the ground. Sometimes she pursues the cruel animal that stings her, sometimes she tries to escape it by fleeing. Then, exclamations reiterated, the most vivid complaints intertwine with the cries of the assistants "AK-AKO," who clap in rhythm with their hands. The beat of the music accelerates more and more. The tambourine and the piercing sound of the castanets accompany the jerky movements of the *almee*. Exhausted with fatigue, the voice faint, one only hears the piercing screams of her companions: "*Ak-ako embeillaho*": Here it is, here's the bee! The torture she endures is unbearable. Finally, she loses all restraint; she gets rid of her veil; her hair falls in disorder on her shoulder; but the cruel animal persists. It continues to pierce her with its stinger. It takes refuge under her belt. She pulls it off violently. He goes under the spincer, under the dress. Same movements, same intensity, everything is taken off; finally, under the pantaloons, the last refuge of modesty. The *almee* brings her hand to it and, torn between two feelings, she abandons this last piece of clothing and faints to the floor. A veil thrown with precision and agility by the matron of the house hides instantly the nudity of the dancer.

The Arab consumes little. His food and his clothing cost but fifteen francs a year. Three small breads of *dura* as wide as three six-pound coins are sufficient for his daily meals. The most industrious add to that watermelons and cucumbers, chicory, a few dates, onions, and his greatest delight, red lentils. In the season of chickpeas, in order to save the *dura*, they drag themselves into the fields and remove a few handfuls of this vegetable, which they eat avidly. The life of the Arab is very monotonous. He rises with the sun and goes to bed shortly after its disappearance. His habitual state, and that which he likes the most, is idleness. At certain times of the year he works, it is true, cultivating the land, but without ardor or zeal. It is not that the Arab is incapable of occasionally showing great activity. One should rather attribute the nonchalance to how little he gets from his labors. In the evening, he returns home, where his women are busy organizing some distractions. He is generally very gay, resigned, good, intelligent, extremely skillful, susceptible to learning, a little irritable, chatty, always ready to do what is asked of him, especially if he is tempted by the lure of the smallest salary. Although impulsive, he bears pains and bad treatment with astonishing patience. It is this last observation, as I have already pointed out, that is the most notable difference between the Arab of the Thebaide and that of Nubia. The latter is more independent and more inclined to avoid violence.

We know how much the Arab likes the sense of the marvelous. The idea of *A Thousand and One Nights* is witness enough, in this respect, to this particular talent of these people, and Egypt has been for all times, and still is today, the country of tales and fables of the supernatural. Each Arab wears an amulet around his neck or around his arm (usually a square of cloth attached with a black ribbon) that he considers as a talisman against the curses of the Maghribi.* This is the name of the very skilled performers who travel through Egypt in all directions, exploiting the credulity of the people. They are supposed to have the ability to sense the presence of serpents and scorpions, and to charm these animals. They claim mainly to predict the future, to see at a distance, or rather to make people see, because here is the

* Translators' note: That is, "the Moroccan."

means that they use to achieve this last result: They choose a young boy, pure of heart, placing in the palm of his hand a wax ring that they fill with ink in order to form a mirror. They then draw on his forehead a few kabbalistic signs, then pronounce meaningless formulae and spray a special perfume in a small dish. Right away, the little boy announces that he sees the sultan or pasha seated on his couch, surrounded by high dignitaries covered with superb diamonds. He then depicts the posture, the clothes, the physiognomy of whoever the spectator asks him about, etc., etc.

These practices did not occur only in Egypt. During the regency,* at a time when philosophy came to replace religious beliefs, people would see in mirrors, or in a glass of water, a whole bunch of extraordinary things. Perhaps this is linked to the effects, more or less proven, more or less debatable, of animal magnetism. Whatever it is, the Maghribis, whose guile and double-dealing have been demonstrated to us more than once, easily impress the Arab population, and even sometimes Europeans, with their skill. The physician of Sherif Bey had so much faith in these performances that he was determined to convince me by trying it himself, in front of me. For this effect he chose, among the 150 children of all ages who worked on the excavation, a small eight-year-old boy, who was submitted to this burlesque experience. The incense was burning, the magical words had been pronounced, and I was waiting for the revelations that were to come out of the mouth of this young oracle. But in vain the doctor repeated the kabbalistic words; in vain he poured handfuls of perfume on the fire. With his eyes fixated on his hand full of ink, the little boy, who had not been taught by the Maghribis, kept saying only that he saw a ring of wax filled with ink, which was the exact truth.

During our stay in Luxor, one of these sorcerers came to pay us a visit. The attempt of the doctor had already given me a rather unfavorable idea of their talent. I asked him to get rid of three snakes who had established themselves under the roof of my house. "Right away," he said, "I am going to make them go into this leather bag." Having positioned himself at the center of the room, like a gladiator, he uttered, or rather screamed, words for a quarter of an hour, mysterious words, while waving a long stick that he held in his hands. Then, all of a sudden, as if he had been seized by convulsive movements and under the control of an invisible agent dragging him in a specific direction, he progressed by bursts, and while briskly waving his arms, toward the window, whose lintel was formed of two rough palm trees. That was the place that he had chosen to furtively place a huge scorpion without his stinger. But he couldn't do it well enough for me not to see it. Caught red-handed, the poor Maghribi was forced to admit his deception. However, I rewarded

* Translators' note: The regency (1715–23) of Philippe d'Orléans for the infant king Louis XV.

him, in order to force him to show us the reptiles he used in his performance. My requests, at first, remained fruitless, but the sight of gold had better success. The Maghribi removed a leather belt containing venomous snakes. Since the officers of the *Luxor* said they were sorry not to see a species for which they would have paid more, the snake charmer could not resist such enticing bait. He unwrapped his turban, in which he had coiled a bunch of reptiles.

In nearly all the villages, there exists a group of low-life women who are very respected by the locals. Their state is regarded as a necessary occupation, a path to arrive in Muhammad's paradise. They welcome European travelers and try to attract them with an agreeable demeanor. They do not live in the same place all the time; they form a sort of nomadic population. One must not believe, however, in spite of the respect that people profess for these women, that the breakdown of mores is a thing tolerated in Egypt, or even seen with indifference. The sex act is carefully monitored, and everyone, even the children, take part in this surveillance. As soon as a stranger comes near an Egyptian woman, she is immediately followed by an Arab, who watches her from a distance. As for the young girls, the loss of their virginity to a European is punished by death. They cut off the head and the body is thrown into the Nile. The guilty man, caught red-handed, is forced, said our interpreter, to become a Muslim if he wants to save the young woman's life. Be that as it may, it is certain that all other dalliances are tolerated and do not cause any danger. All the women nurse their children themselves; it is a law of nature. It is violated only in civilized countries where the citizens are divided into classes with respect to their wealth. They pay with gold for the pleasure that they get from the work of the lower classes. Here, there is no distinction; all are equal. This equality does not arise from the exercise of human rights and from an equality of social well-being, as it does in the United States, for example, but it comes from the equality of slavery. Everything here is slavery. The sword that travels horizontally has leveled all heads. And if one finds a few exceptions, if a few individuals manage to create a superior condition, it is only under conditions that I do not dare tell. Most of the Turks who impose them were themselves subjected to these conditions from their youth, in order to occupy the position that they have reached.

The periodic flooding of the Nile seems to cause a general regularity in the habits of the animals and even of physical phenomena. The departure and return of the birds, the regeneration of insects, are coordinated with the rising and lowering of the Nile. The pigeons, which are extremely numerous in Egypt, leave the coop to search for food in the countryside, leave the fields to drink in the Nile, come back home, take their nap during the greatest heat of the day, and return again to the countryside and the banks of the Nile, always together and at the same time of day. The buffaloes and the dogs also go at a specific time of day to take a bath in the river. The latter animal, although there as everywhere the friend of man, does not live inside of houses. He only goes inside very rarely and always finds his place nearby. This habit seems innate to him. The sailors of the expedition, who had taken in a newborn dog, raised him and fed him inside the temple. After five months, having been let free, the animal went right away to settle outside like other dogs of the country. He had to be brought back forcefully. Several other trials were attempted whose results were the same: As soon as he was let go, he would take up his position in the surrounding outdoors.

The strength of the winds follows incrementally the rising or lowering of the waters. Those of the north and northwest predominate for the entire duration of the inundation. Their intensity is almost proportional to the increase of the speed of the current. This combination allows the boats to go up and down the river at all times, using the sails or dropping them.*

The south winds are also periodic. They blow several times during a period of fifty days, which gives them the name of *khamasin*, the Arabic word that means fifty.

The immutability in the various phenomena of which we just spoke in the state of the atmosphere, which, except for a few days per year, constantly keeps the most perfect purity, inspires in the Egyptians this sense of order, of regularity, of grandeur, that characterizes all their works. The same reason explains the excessive repugnance that they feel toward all innovation, for everything that is foreign to their customs, their way of life, their laws. This is perhaps one of the causes which, while keeping the Egyptian nation from assimilating the ways of life, the customs, and the ideas of the people who invaded it successively, have contributed to keep it in a perpetual slavery.

During the season of low water, the heat is excessive and the *khamasin*, which very fortunately only blows intermittently, causes very dangerous diseases. This impetuous wind is always preceded by specific signs. The air, the birds, and the animals announce by their frantic actions impending disaster. The beautiful sky of the Thebaid, usually so peaceful, is suddenly darkened by clouds of dust which, divided into its smallest elements, covers and penetrates every object. The sands of the desert, lifted like an angry sea, inundate

* Herodotus describes, in these terms, the way the boats were handled while going down the Nile: "They have a raft made of tamarisk wood, fastened together with matting of reeds, and a pierced stone of about two talents' weight; the raft is let go to float down ahead of the boat, made fast to it by a rope, and the stone is made fast also by a rope to the after part of the boat. So, driven by the current, the raft floats swiftly and tows the 'baris' (which is the name of these boats), and the stone dragging behind on the river bottom keeps the boat's course straight."

Today, we just turn the two sides of the boat alternately to the wind, which by its action on the rigging and the hull produces the same effect as the stone of which Herodotus spoke. If the weather is calm, or if the wind is favorable, the oars are used to steer or to go faster than the current.

the cultivatable lands and spread their desolation. The atmosphere, heavy with heat, dries the skin, oppresses the lungs, irritates the nerves, and lingers in the throat. The sands are burning hot. You blow your nose and spit out the sand. Any sense of pleasure abandons you, weakness takes over your limbs and provokes a general malaise, a melancholy that weighs on the spirit and consumes the principle of life. The expansion of the air is so great that multiple inhalings are insufficient to fill the lungs. Then you resort to a short dry panting, involuntary movements, a continual change of place to find a bit of air less heated and less deprived of oxygen. A man placed at the entrance of an oven will only get an imperfect idea of the sensation caused by the *khamasin*. This painful state ends as soon as the wind takes another direction. As the Arabs say, the unveiled sun drinks up the ignited particles that swirl in space, and the sands, forming furious waves, retreat into the desert. Perspiration, a certain sign of great health, returns, and the vital forces regain part of their energy. But the influence of the climate is always there, undermining everyone stealthily, and a longer stay in Luxor would have caused the death of a great number of our men. They had become pale, weak, and listless. One cannot become acclimated to this leaden sky; it is necessary to be of Arab descent to breathe freely this air of fire. The son of a European and a native woman rarely reaches his tenth year. The Mamluks themselves, who lived in and governed the Said, have not left any descendants. On the other hand, children of the pure race resist this excessive heat perfectly and without precautions. One sees them in temperatures of forty degrees, naked, their heads uncovered, playing, having fun, running, jumping in the river, then going back to their fun, rolling in the sand, without their health ever being altered.

The illnesses that particularly attack Europeans are ophthalmia and dysentery. The effects of the latter are terrible and the only remedy against it is a change of climate.

Before my departure for Nubia, I had hastily visited the surroundings of Luxor. Upon returning, I continued my investigations. I saw again, with the same feeling of admiration and astonishment, the ruins of Karnak, which still, to a poetic imagination, bring to mind the queen of cities, the city of marvels. An irresistible attraction pushes you toward these masterpieces of art. They rise on a constructed hill, situated in the middle of a large cultivable plain. Temples, palaces, gigantic masses, series of monumental columns, whose harmonious architecture one could never tire of admiring—colossi, obelisks, whose carving is of a purity one could call unparalleled. Such is the beautiful development of greatness and magnificence afforded to your eyes, which electrifies the least impressionable man. No people has understood as well as the Egyptians this magic power of art acting on the soul through the senses, this progression, one could say this dramatic appeal, that ravishes the spectator. It is impossible to believe, even after having witnessed the evidence of so many buildings in the same place. Their description seems a fantasy. It resembles a dream of the *Thousand and One Nights*, a fairy tale.

Built of stone, imposing because of their severe form that is found nowhere else, these edifices are also remarkable for their excellence of execution required by their grand proportions. They bear, in addition, a primitive character, the mark of their purpose, and of an architecture that takes everything from pure imagination and from the ground where it was born. The arrangement of the plan is simple, easy to conceive; they do not belong to those imaginatively conceived combinations that tire the observer without pleasing him. All is clear, neat, and precise, intelligible to everyone. The eyes rest agreeably on forests of columns whose numbers and proportions are in harmony with the soaring ceilings they support. The slope, principle of solidity, is observed everywhere; all the great masses rising in the pyramid shape are bounded by a torus that seems to encircle them. But what astonishes above all is the magical effect of the small number of elements that compose this architecture. Always thought out, everything in accord, well planned, the architectural lines present greatness and purity; perfectly straight edges surround the entire edifice without interruption. They are never cut or altered by other ornaments, which look rich from close up without ruining the general effect from afar. It is only from a short distance that the eye can distinguish the multiplicity of interesting and educational details that cover these monuments. The walls, the ceilings, and even the tiniest nook, inside or outside, all are covered with hieroglyphic tableaux, accompanied by an explanation that is itself a decoration. These tableaux, already so amazing by their variety, richness, and vivacity of the colors, capture the attention with an interest constantly renewed. These are huge historical pages where ethics and science are developed, where all arts are put forward. One can see religious ceremonies, altars full of offerings, battles on land and on the sea, the crossing of a river by an army, the taking of a fortress, the trophies of the victor, the tributes levied on the vanquished, complete systems of astronomy, all the produce of the land, of the arts, of commerce and industry, the customs of the most ancient civilized people. These representations, sculpted in relief, are combined with a marvelous ease to teach us and show us in action the arms, the chariots, the instruments of war, the most memorable characteristics of the combatants, the misfortune of the defeated and the way in which they are treated, the honors given to the victors, etc., etc., etc. All the figures, with the exception of those on the sacred bas-reliefs, where stiffness and immobility are constantly observed, express through their attitudes the various emotions that animate them. The suffering is most well rendered, and everything, indeed, has been calculated, set, to produce the deepest feelings. This is what gives them such a powerful charm, so that one cannot leave these ruins without the desire to see them again.

The mysterious characters that cover these majestic relics present a new attraction for the curiosity of the scholar and archaeologist. On all their surfaces is engraved, in a language almost unknown, the only history of the time when they were erected, and what times! Those when Genesis had not yet been written, when Moses was leading the tribes of Israel through the desert (the obelisk of Paris is contemporary to that time), when everything that we call antiquity—Homer, Lycurgus, Alexander—hadn't yet been born. For this last reason, they are especially precious as historical monuments, since through erudite and judicious studies Champollion the Younger has succeeded in finding the meaning of hieroglyphic writing. But how many riches are still buried in this soil that one needs merely to dig in order to bring to the light of day new discoveries about this great people! The clearing of a tomb, of which one could see only the cornice, showed us on the four sides all the arts of Europe sculpted in various colors with perfection of truly astonishing details: the glass blowers, sculptors, carpenters, shoemakers, farming instruments, lyres, etched vessels, etc., etc., etc. This tomb is located on the other side of the river near the Valley of Biban al-Muluk.* Here the Egyptians have dug an entire chain of mountains to house in magnificence the mummies of twenty-three pharaohs. The surroundings of the underground palaces attest to the great principle of arrangement and harmony from which the Egyptians never went astray in all their conceptions, comparing them to these immutable laws that rule over matter, without which all things would fall into disorder and confusion. These underground tombs are dug deep in the rock itself, around a valley where nothing can be grown, and are surrounded by inaccessible mountains. One can only penetrate this enclosure through a man-made opening. Everything here presents a more arid aspect, wilder and more isolated than the most terrible deserts. Here the temperature is sometimes cooled by the northwest breeze; then the air is always calm. The winds bounce off the belt of sandstone that surrounds the valley. The heat concentrates there and becomes extreme, unbearable at the moment when the sun throws its rays at the bottom of the valley. One walks on burning sands, one breathes under a sky of fire, the feeling of life abandons one in the middle of silence, the calm, the perfect stillness, the nothingness of everything that surrounds the resting place of the dead. Dominated by these impressions, the traveler goes towards a vertical door cut within the rock. It is the entrance to the tomb of the pharaoh. By torchlight, one walks along a long gallery that runs along the inside of the mountain, always on an incline. One finds from place to place doorframes cut in the rock intended to accommodate doors. This gallery leads to square or oblong rooms whose pillars support the mountain. One of them, the largest of all, usually contains a beautiful

* Translators' note: The Valley of the Kings.

sarcophagus. But, if there is anything surprising in the world, it is the prodigious quantity of paintings that decorate so richly all the walls of these somber dwellings. They represent systems of astronomy, all of Egyptian mythology, the most important acts of politics, legislation, worship and customs, ceremonies of civil life, marriages, funerals, diverse domestic occupations, the products of the arts and industry, etc., etc., etc. These carvings, which occupy more than a square league of surface, are painted and worked with a purity and precision that surprise the artist. One notices even the most detailed aspect of the fur of the animals. It is unbelievable, I repeat, even after having seen it. One is astounded at the thought of such a project whose execution must have cost such pain, such work, running up incalculable expenses and demanding a dedication, a superhuman obsession.

These diversions dissipated a little the melancholy that permeated us. The days are so long and so sad when one is suffering a thousand leagues from one's country, and without any news. One's only pleasure was to visit the Valley of Biban al-Muluk and the ruins of Thebes. At night, each would tell what he had seen, the emotions he had experienced, or bring back new cartouches* whose name we tried to figure out. But always, the conversation would end with this phrase: *The Nile has gone down this many more centimeters*, and all eyes would sadly turn to the ship, run aground on the bank, and resembling more a mountain of sand than a boat destined to transport us to France.

Before the rise of the river, it was decided to go with the captain of the *Luxor* to Rosetta in order to examine the state of the *boghas*, and to see if it was necessary to build pontoons† in order to facilitate the passage over the bar. This trip showing nothing unusual worth noting, I will just say a few words about our meeting with the pasha.

Muhammad Ali received me with the strongest show of satisfaction, and inquired about the happy ending of our enterprise, praises that it was easy for me to reciprocate, complimenting him on the taking of St. Jean of Acre by his son, Ibrahim. Muhammad Ali had received the news of this victory the day before. "I am assured," he said to me, "of the conquest of Syria. I examined the map of the Ottoman Empire for five months, night and day. I figured the numbers of troops that might oppose me, and these calculations showed me the sufficiency of my army over those I would have to combat." He gave me explanations on this subject that seemed to announce great ambitions.

His Highness then asked me several questions on the nautical qualities of his ships and on naval tactics.

* Hieroglyphic inscriptions set in a specific way, mentioning the name of a king or queen. One recognizes these cartouches by the oval that always surrounds the inscription.

† Transport ships intended to lift out of the water other ships drawing more water than they.

"I don't want," he told me, "to join in a traditional sea battle. My admirals received the order to watch the enemy's flotilla and, once it started moving, to destroy the first ship that found itself separated from its squadron by a false maneuver. I know them," he added. "The capture or destruction of one of their ships will suffice to set the entire army running." Bringing the conversation back to the obelisk, he asked me how much of its weight it would lose in the water. After I had answered his question, he replied: "I remember in my youth, I would carry stones under water for a great distance that I could not have moved on land. Such as you see me, I was destined to be an engineer." To which I answered, jokingly, "If Your Highness so desires, we will change roles."

On our return to Luxor, several signs of rising began to appear. The boat was entirely rid of the sand enveloping it, which we wet two times a day to protect it from the heat of the sun. This manner of preservation succeeded perfectly. The cladding of the ship was found to be unchanged, which one could see through the seams of the keelson. The silt of the river, deposited by the last inundation between the five keels, was still damp, although it had not been reached by the water thrown twice daily. It appears that these alluvials behave like clays. Exposed to a very high temperature, they strongly retain the water particles that they have absorbed.

Finally, around the first of June, we started to notice the rising of the Nile. The waters lose their transparency; waves begin to be seen. With this news, excitement was among us. We ran toward the river, we looked, we examined; we placed indicators. Happiness and hope animated all faces. The new height was published at the door of the mosque, and the Arab women celebrated it in the evening on the banks of the river with religious songs. The following days, until July 30, the water rose quite rapidly. The water was spreading out more and more on both banks, and one thought absorbed us night and day. Every hour, every minute, we would visit the Nilometer. A few, more impatient to see the mother country again, and tormented by a horrible uncertainty, would spend part of the night examining the progress of the Nile. It was a curious sight to see them lean over the banks, searching blindly for the previously installed markers, to rejoice or become sad according to the results of their observations. Finally, on August 5, the ship formed an island that only needed a rise of thirty centimeters to float. Tomorrow, or the day after, at the latest, the problem would be resolved, the ship would float. But on August 6, beginning in the morning, the Nile started to go down, and continued the downward movement till August 12. This event produced in all the spirits a kind of stupor, and caused the greatest worries for several days. We asked all the Arabs if they had witnessed such a plague, or if they hoped that the Nile would start to rise again. The oldest priest in the mosque, interrogated by us, answered our question with great seriousness: "Child, do not fear anything! Muhammad is good and great and the land is not inundated yet.

The Nile, like a spirited horse who has just run a long course, pants to get his breath back.* In a short time, it will surprise you with its new vigor. Meanwhile, make an offering of two sheep to the mosque." The prediction of the venerable one soon came true. On the 18th, the ship, carrying its precious load and floating on the most beautiful and peaceful river, was launched in the direction of the current.

The departure of the *Luxor* took place on August 25. Its navigation on the river presented nothing remarkable. Two Arab pilots were charged with signaling the sand bars and the direction of the current. We proceeded in very short stages. We would stop every evening. We arrived in Rosetta on October 1. Twelve days earlier, the bar in the river would have been crossed without any difficulty, but now the passage was impossible. We waited in vain during the entire month of October. I thus had to find a way to lighten the ship in order to decrease the amount of water that it drew. This ship, built in pinewood, and not a very strong example, did not have the strength necessary to withstand the push of two pontoons. All the means used and known in such a case were inapplicable.

Plate VI gives a sufficient idea of the apparatus that we proposed to use. It consisted of two pontoons placed on each side of the *Luxor*, which were meant to act on the obelisk in the same way that dockhands transport a load attached to several bars whose ends they lift with their shoulders. By this means, the ship would not feel any pressure and the water that it was drawing would be diminished in proportion to the force exerted by the pontoons on the mass.

This operation could only be done during the fair-weather season, when the sea is calm. The pilot of the *boghas*, losing hope of our crossing the sandbar before this time, encouraged us to stay at Rosetta, where the ship was completely dismasted.

Our sojourn in that town would have been prolonged till April had not an accident happened to a boat from Anatolia carrying oranges. The pilot, having been called to right that ship, announced that, in turn, a passage had formed in the sand bar. Immediately everyone set to work rigging the boat. This took three days of work. The first of January at eleven at night, the *Luxor*, after having grounded several times, finally entered the sea. The pilot's maneuver, helped by his four children, deserved the greatest praise and proves what I have already said elsewhere about the faculty of observation and the precision of execution that distinguish the Arab.

* The language of the Arabs is constantly filled with metaphors and images, with more or less pompous and exaggerated comparisons. In the same way that the priest of the mosque compared the upward movement of the Nile to a spirited horse, the *nazir* of Luxor, speaking of the plain of Thebes after the removal of the waters, said to us: "It is an emerald shining with the sharpest colors"; and his lieutenant, seeing a well-formed young lady, would cry out: "See this young and beautiful palm tree, walking along bearing all its fruits."

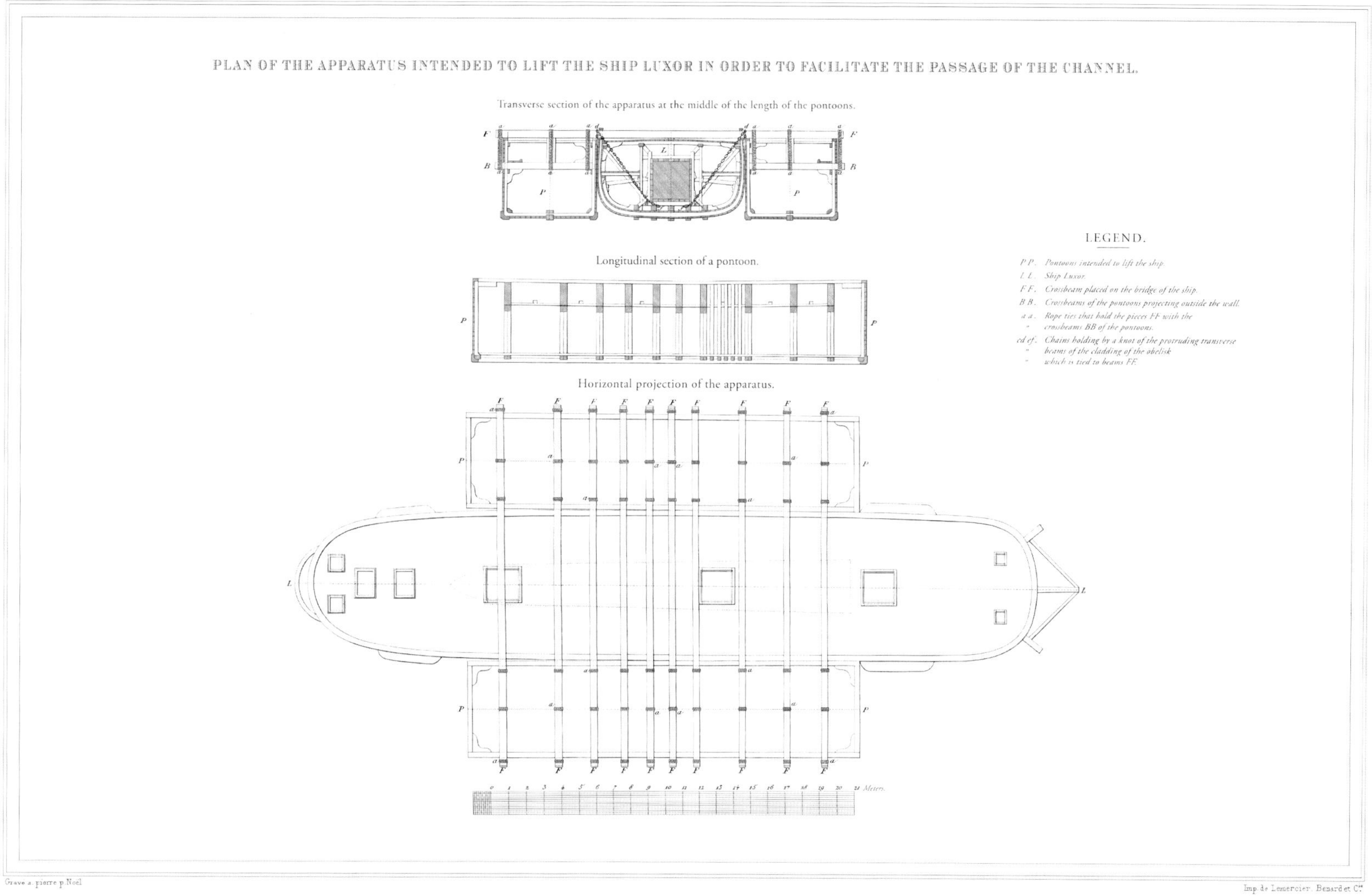

Plate VI: Plan of the apparatus intended to lift the ship *Luxor* in order to facilitate the passage of the channel

The steamship *Sphinx* (right) towing the *Luxor* to France in high seas, drawn by Léon de Johannis, second-in-command on the *Luxor*

At midnight, the *Sphinx* took the *Luxor* in tow and took it to Alexandria, where it moored on January 2 at 9:00 a.m. The variability and unpredictability of the winds detained us in the port till April 1. On that day, we set sail to return to France with a favorable breeze. The good weather only lasted a day, and soon contrary winds and a rolling sea forced us to shelter in Rhodes, Marmara, Milo, Navarin, Zantes, and Corfu. It was in this last port that the steamboat was able to replenish its depleted coal. We left Corfu on May 2 and, continuing on our way, towed by the *Sphinx*, we soon could make out the coasts of Provence. Finally, the *Luxor* dropped anchor in the harbor of Toulon in the night from May 10 to 11, 1833, bringing in its hull, to French soil, the precious monolith, witness to the glory of the ancient pharaohs.

PART 3 — Work in Paris

Departure of Mr. Lebas for Paris — He is charged by Mr. Thiers with the erection of the obelisk — Construction of the docking cradle — Putting the apparatus in place and hauling over land — First moving of the obelisk: passage on the bridge ramp — Second moving: hauling to the Place de la Concorde — Third moving: hauling up to the viaduct — Fourth moving: the obelisk arrives at the level of its pedestal — Setting up the erection equipment — Erection of the obelisk

After a month of quarantine, the *Luxor* entered the dockyard of Toulon. The following day, it anchored in its berth, where it was only supposed to stay for a few days.

The boat was inspected and we observed that its hull was in a good state of conservation. No loosening was noticed at the cut where the obelisk was placed in the ship. The long and laborious navigation of the *Luxor* on the Nile and the Mediterranean was, in itself, a sufficient guarantee for the rest of the route it would have to traverse. We could hope with confidence that by leaving Toulon on June 22, it would arrive toward the end of August at Le Havre. There was no doubt that we would take advantage of the first rise of the Seine to bring it to Paris.

During the quarantine, I had received the order to go to the capital as soon as we were released.

Mr. Thiers, who lent the arts and sciences an enlightened protection, was then minister of commerce and public works. Once the obelisk arrived at Paris, it came under the jurisdiction of his department and my mission was thus finished. But the minister decided that the honor of erecting the monument should be given to the one who had taken it down and loaded it on board the *Luxor* in the sands of the Thebaid, which was only fair. On August 10, 1833, I received the following notification:

> Sir,
>
> I have the honor to tell you that the Minister of Commerce and Public Works has decided that you would be put in charge of works concerning unloading, transport to the site, and the erection of the obelisk of Luxor on its pedestal.

The skill of which you have given so many proofs, by bringing the obelisk all the way to France, guarantees the complete success of the new operation that you will undertake.

Please receive, sir, the assurance of my distinguished respect.

The master of requisitions charged with the direction of civil constructions,

Signed, Edmond Blanc

The operations on the obelisk would be numerous and complicated, and would demand a lot of time. A few could only be carried out in the season of low water. It would have been dangerous to dock a boat of sapwood [pine] on a rough surface, with one or several projections. It was absolutely necessary that the five keels should rest on strong supports, perfectly regular, in order to prevent the rupture of the hull, and possibly the monolith. A docking cradle would thus be indispensable. Its position on the bank of the Seine would depend on the choice of the place where the obelisk would be erected. Everything was, of course, put on hold until this last choice was made.

The center of the Place de la Concorde, successively occupied by various monuments, had been, under the Restoration, dedicated to the *Apotheosis of Louis XVI*. After the revolution of 1830, they decided to substitute for it one of the monuments of Sesostris, which was on its way to the capital across the vast ocean.

The bottom of the right ramp of the Pont de la Concorde, the point nearest the site, was chosen to dock the *Luxor*. We began immediately to remove the debris that encumbered this part of the bank. Meanwhile, the carpenters were preparing the docking cradle.

The cradle was composed of cross-struts placed a meter apart and tied between them lengthwise by five external bands, corresponding to the five keels of the ship (Plates VII and VIII). Its level had to be calculated so that the *Luxor* would be able to float on top of it when the depth of the river allowed the ship to come from Rouen to Paris.

When the ground had been leveled to the low-water mark of the river, we intended to set it following the slope of the ramp, but the delays caused by a strike by the carpenters, who were seeking a raise, prevented giving it that slope. During the suspension of work, the Seine rose several feet; we couldn't think of digging in flooded soil, much less to set up the frame on dry land. Some workers immersed in the water up to their armpits assembled all the parts of the cradle on a flat raft held in place by ropes fixed to pilings. Then we filled in with pebbles the empty spaces between the longitudinal beams and the braces between the beams in order to sink the system to the bottom of the water, which was done by easing off on the retaining cables. It was there that the *Luxor* came to rest on December 23, 1833.

Once the cradle was completed in October, my first task was to draw a general plan of the site where we had to operate, and to combine the means that I intended to use to unload the obelisk, haul it, and place it on its base.

The large area of the land did not allow me to draw the plan on a large scale. I tried, however, to represent as clearly as possible the whole of the operation. The explanations I am going to give will supplement the details that are not present in the plan.

The work to be done was of two kinds. It was first necessary to raise the obelisk to the top of the pedestal without changing its inclination, then to make it turn a quarter circle while it was suspended, to raise it onto the resting bed, whose dimensions were exactly the same as those of the base of the monolith.

Plates VII, VIII, IX, and X can give an exact idea of the length and the various slopes of the path that the monolith had to cover.

The first part, thirty meters long, was almost horizontal and ended at the bottom of the ramp, with a six-degree slope. Beyond, there was a path of different slope, deviating at A to avoid the angle of the ditch, F. This path ended at point B, at the beginning of the square on the side of the bridge. Having arrived at this position, the obelisk had to turn at a right angle and then go onto a viaduct, which rose from that point following a slope of eight centimeters per meter until the top of the pedestal. There ended the first part of the work, which is, properly speaking, the transport.

In August 1834, the water having left the *Luxor* high and dry, we proceeded with the demolition of the front of the ship and with the construction of the sliding cradle.

This cradle was composed of two longitudinal beams placed on an oak floor built from the opening of the ship up to the height of the quay, following the irregularities of the land (Plate VIII). Thus it was first necessary to drag the monolith to CC, the intersection of the two roads. Then, to execute the double movement of the monument, we made it go up on the embankment of the ramp.

To accomplish this we built a launching cradle, abC, whose top served as an extension of the horizontal road, or rather the keelson of the ship. This cradle, intended to carry the monolith in its upward journey, rested on the longitudinal beams of the loading dock and was supposed to go up, sliding on these pieces of wood.

Its placement demanded that we excavate a trench the same length as the obelisk, following the slope of the ramp and starting from its intersection with the upper plane of the ship's hull struts LL; but the muddy terrain, diluted by the water surging from the bottom, did not allow us to extend it beyond 15.5 meters. Having reached this distance, it was necessary to remove the water from the excavation in order to clean up and reinforce the ground. I recoiled from the expenses that such a work would have cost. It was decided that the cradle would be reduced 5.5 meters lengthwise so that the monolith would stick out the same amount from this construction, which was intended to support it in its movement up the inclined plane.

Plate VII: Perspective view of the path followed by the obelisk

LLOWED BY THE OBELISK.

Imp. de Lemercier, Benard et Cie.

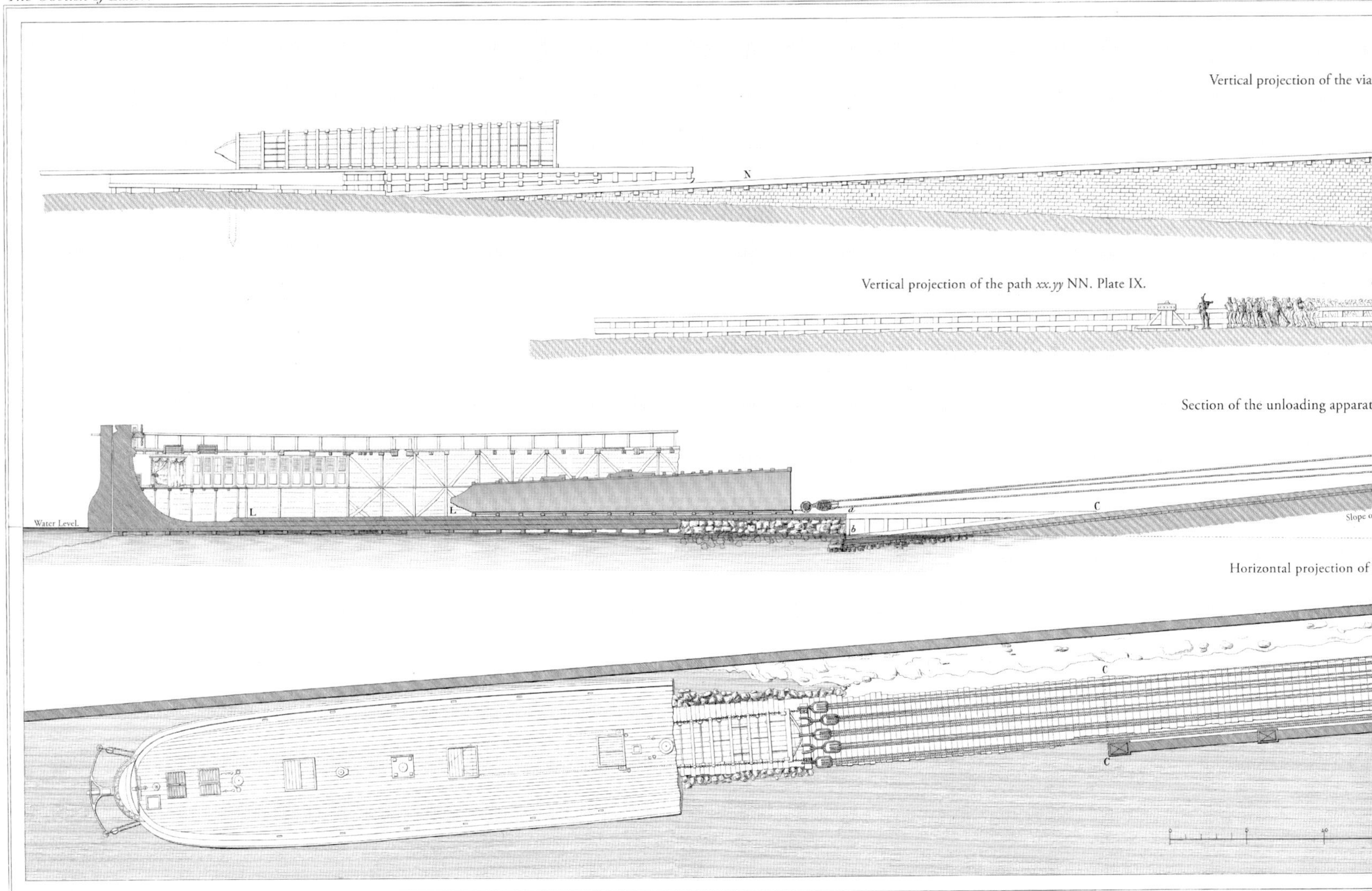

Gravé sur pierre p. Noel.

Plate VIII: Stages of disembarking the obelisk

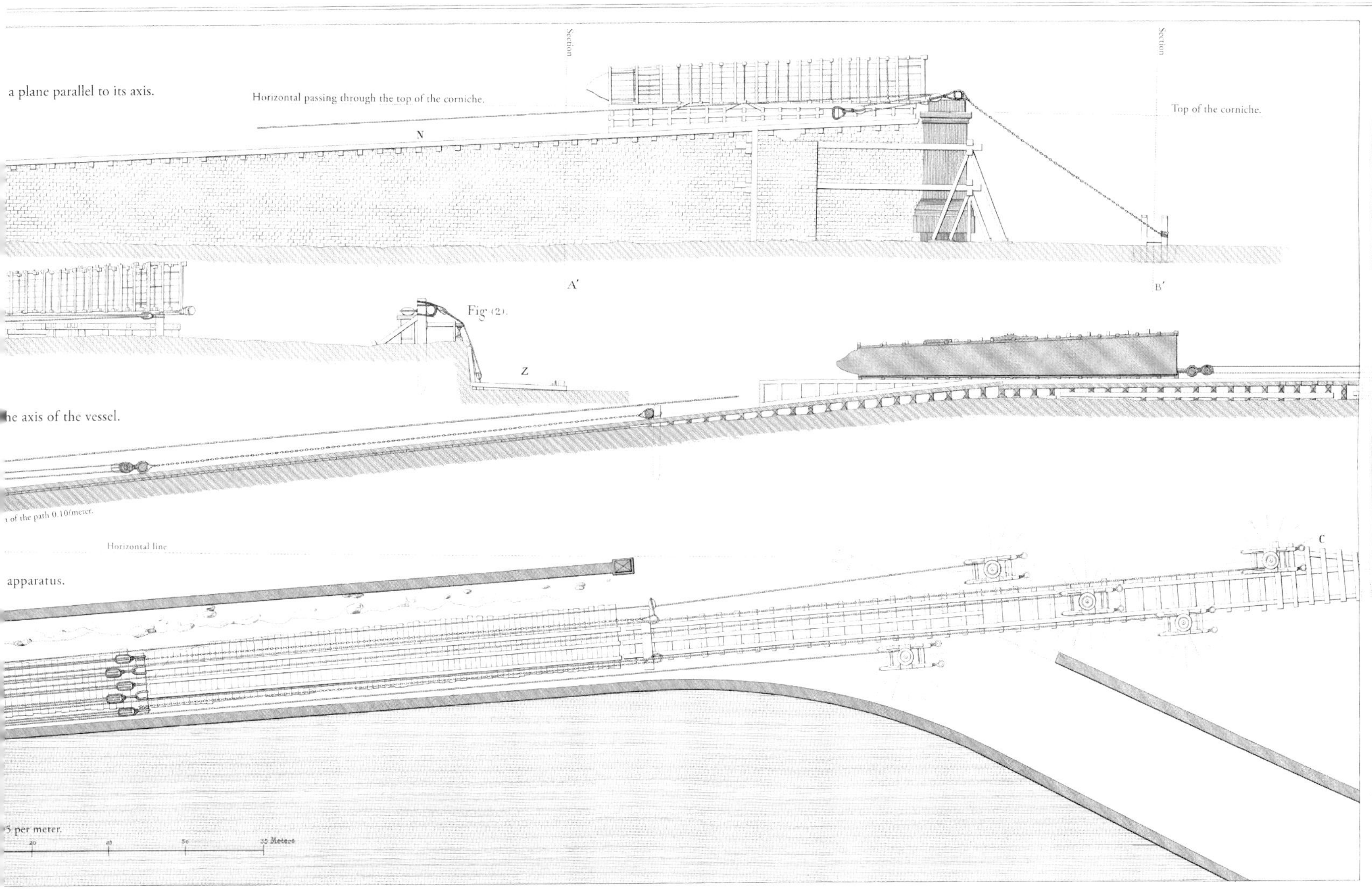

Imp: Lemercier, Bénard et Cie

Horizontal Projection of Part of the Path Taken by the Obelisk.

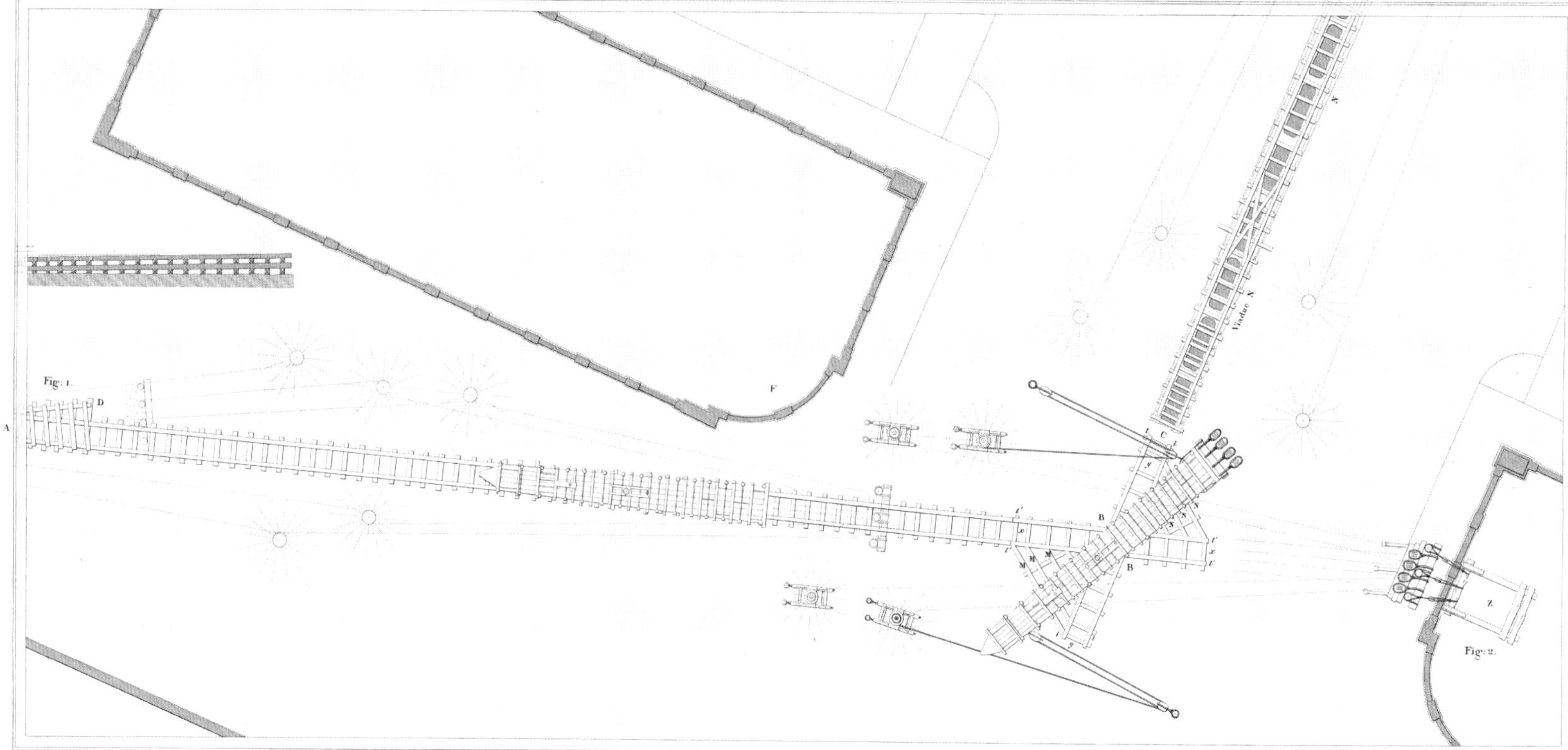

Gravé s. pierre p. Noël

Imp. de Lemercier Bénard et Cie

Plate IX: Horizontal projection of part of the path taken by the obelisk

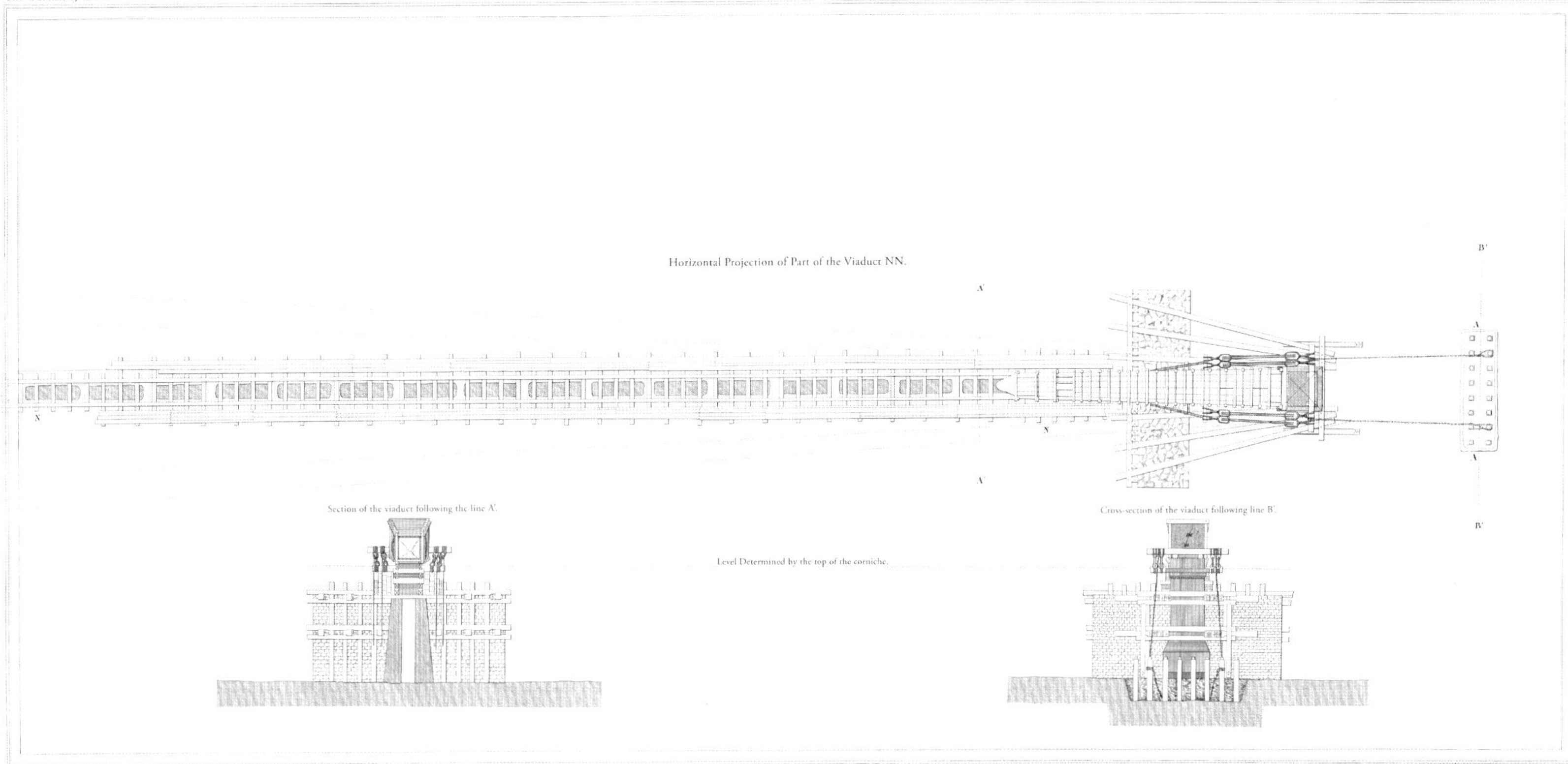

Plate X: Detail of the route of the obelisk

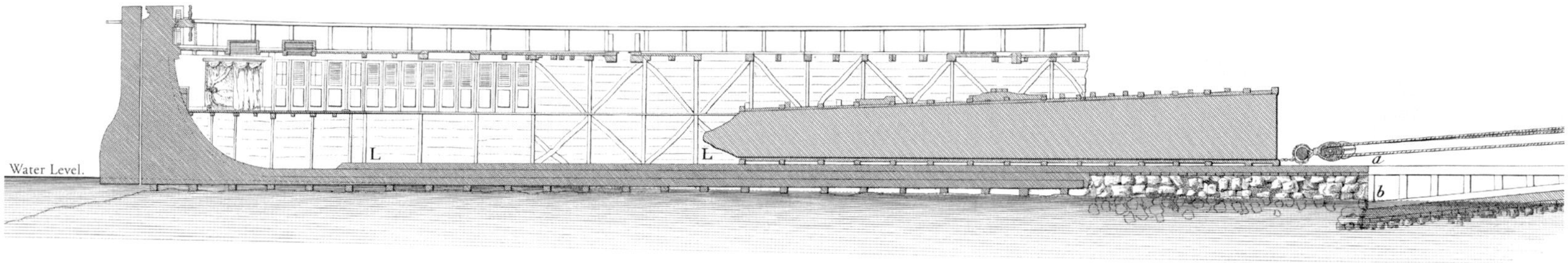

Detail from Plate VIII: The launching cradle

With the help of the cradle, whose angle varied because of the fluctuations of the ground, we succeeded in moving the obelisk to the top of the pedestal while always keeping it in the horizontal position that it occupied on board the ship.

Setting Up the System and Hauling on Land (Plate VIII)

The anchor chain of the ship was placed as a belt around the obelisk. Its two extremities were fixed to a roller one meter from the base. The roller carried five pulleys with six branches. The fixed pulleys of these derricks were attached to another roller held by chains whose ends were attached to anchored piles at the head of the ramp.

Five capstans were arranged in a checkerboard pattern on the road of the Quai Conférence. The gate had been demolished to allow free passage to the heads of the tackle that wound themselves around the bells of these motors. Each capstan, attached to two piles, was provided with sixteen bars, upon which forty-eight men would push.

The true tension transmitted by each of them to the corresponding pulley would yield 9,510 kilograms, for a force of ten kilograms exerted by each man on the ends of the levers. Since each of them could produce, if need be, a force of twenty kilograms, the five capstans would pull the obelisk as would a weight of 95,100 kg, whereas the maximum resistance to overcome would only be 52,000 kilograms.

It would appear, at first, that the number of apparatuses could have been reduced, but this excess in the intensity of the machinery is more appearance than

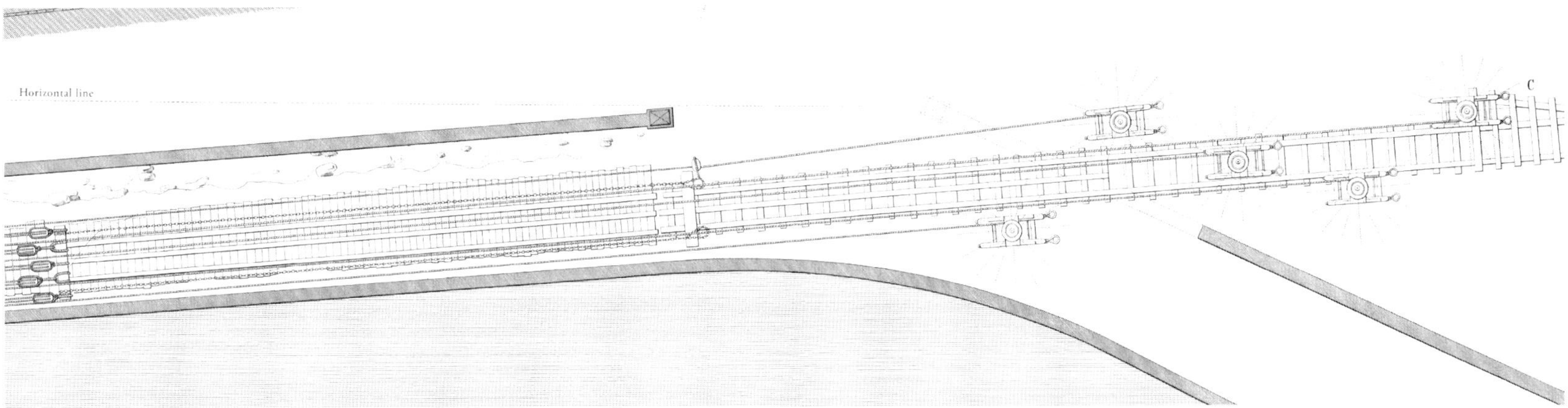

Detail from Plate VIII: Five pulleys attached to five capstans

reality. These results, deduced theoretically, assume that the surfaces of contact are perfectly smooth and well greased, that they offer equal resistance throughout their entire length, that there is neither compression in the wood nor settling in the soil, etc. But experience shows that this is not the case when working on heavy loads that occupy little space relative to their weights. The system deforms, to a greater or lesser degree, under the pressure exerted by the mass. From that fact arise resistances much greater than those that we took into account in the calculations. They are often produced by causes that escape the sharpest eye. This is why, following a slight compression in the wood, the extremity of the cradle removed part or all of the tallow with which we had coated the long beams of the cradle. Thus, the friction becomes double, triple, quadruple what it would have been had the grease placed between the rubbing surfaces not been removed.*

One must thus be prepared to use an excess of force to overcome these unforeseen obstacles; one must overcome them quickly and promptly, because the more you delay, the more difficult the task will become.

First Moving of the Monolith (Plate VIII)

On August 9 all was ready, as we have just indicated. The capstans were put in action by 240 artillerymen.

* We tried to partly remedy this problem by the following method. Several funnels were placed on the faces of the long beams of the cradle. The opening of each one was fitted into a horizontal opening, pierced in the lower angle of these pieces in such a way that, by pouring boiling tallow into these reservoirs, the material penetrated between the surfaces in contact as the cradle advanced.

The monolith, yielding to their combined efforts, advanced by little spurts toward the ramp and placed itself on the cradle. In this position, its lower edge coming into contact with the long beams of the dock, progressive horizontal movement became impossible. The motive force now had to be applied, not on the obelisk, but on the cradle. To this effect, the roller holding the mobile pulleys was detached from the belt and fixed to the cradle with a chain wrapped twice between two of its crossbeams. It was supposed to go up, sliding up to the height of the quay, when the force would take over the sum of the resistance due to friction and the (natural) force of the monolith to go down the inclined plane.

The following day, the hauling recommenced. At four o'clock in the evening, the cradle carrying the obelisk had reached the top of the ramp.

It was only after this operation that the dimensions of the blocks to be used in the construction of the pedestal were decided. They were to be extracted from the quarries of L'Aber Ildut.*

The original project called for twenty-seven pieces, but in the course of quarrying, the contractor, having discovered a rock in the harbor of L'Aber itself that would provide a monolith 4.8 meters in length and over 2.0 meters wide, the number of pieces was reduced to five. Their weight was equal to 240 tons or 240,000 kg.

* Translators' note: In Brittany.

The boats *Luxor* and *Sphinx* were used to transport the granite blocks, and I was put in charge of loading the blocks onto the first of these ships.

The operation took place on September 5, 1836. The five blocks of the pedestal, aligned in the order in which they were to be placed, on a single cradle, as if they were one mass, were brought to the same place in the hull of the *Luxor* that the obelisk had occupied; the process took fifty-five minutes.

The next day, six large platforms, weighing 70,000 kg, completed the load, and were brought aboard the same ship.

On October 7 the *Luxor*, whose front had been taken off, put back, and strengthened by the same means used in Egypt, started off, towed by the *Sphinx*.

On December 15 it was anchored in Paris below the Pont de la Concorde Bridge.

Second Movement of the Monolith (Plates VIII and IX)

After the operation of August 1834, whose purpose was to remove the obelisk from the ship that brought it from Thebes, the monolith, placed on its inclined cradle, had remained at the top of the dock of the Pont de la Concorde. Because the lowering of the water would prevent the unloading of the *Luxor* in the month of April, it became urgent to clear the ramp and to haul the obelisk to another part of the quay to allow the transport of the granite blocks from Brittany.

The sliding cradle, placed on the uneven terrain, was divided into two parts located on two different levels, as in the first operation. Thus the movement that had to be carried out consisted first in sliding the cradle supporting the obelisk until it met a second cradle set to go from the slope of the old path to that of the new one. The next step was to transport the monument on the second cradle, which in its turn had to progress up to the point where the whole system was to be placed temporarily.

The carpentry work was started on March 25. On April 16 at 4:00 p.m. the monolith, pulled by four capstans, had left its former cradle and passed onto the other one. We delayed to the following day the rest of the operation, whose most difficult part still remained. For the first time we would need to change the direction of the obelisk, to change its axis.

Figure 1, Plates VIII and IX, gives a sufficiently exact idea of the carpentry system built in D and C. It was on those two pieces of equipment that the obelisk had to slide in order to be placed on the axis of the new path.

The maneuver was executed in the following manner: Some props or stanchions placed to meet at a point at the head of the monolith tended to push this part toward the Champs-Élysées, while the other end was moving toward the Place de la Concorde. A pulley block attached to the base and whose top was tied to a capstan moderated this movement at will, and four capstans, moving all together or successively, caused this imposing mass to advance, to turn, to straighten up, to stop or go, according to the need. At 11:00 the obelisk was found to be perfectly oriented. Three hours later it had arrived at the end of its journey—in other words, at the corner of the ditch next to the bridge.

In this position, the monolith didn't hinder the traffic, and allowed free passage for the transport of the granite blocks destined for the construction of the pedestal. The hauling was thus suspended, while we proceeded to unload and stack these blocks.

Third Movement of the Monolith (Plates VIII and IX)

On August 1, the work was begun again with great activity. We extended the wooden cradle upon which the obelisk rested. At the same time, the masons were raising the viaduct that ended in the center of the square, in such a way that the highest part of this inclined plane was at the exact height of the level of the pedestal. Masonry was preferable to a wooden frame as being more economical, more reliable, and occupying less space on the public road.

The plates to which we refer show that, after having dragged the monument up to the intersection of the two paths, we had to make it turn ninety degrees toward the axis of the bridge. The same drawing clearly indicates the carpentry system used in order to reach that goal. It consisted of a mobile cradle around a stake driven in at the intersection of the axes of the two cradles. This cradle, whose top surface served as a continuation of the road carrying the obelisk, rested on a horizontal platform. It was composed of four long beams, *ll* and *l'l'*, and of several rows of beams placed in the two angles opposite the tip that the long beams formed among themselves, in such a way that the cradle, passing from position *x* to *y*, had to be supported from one location to the other by the intermediate pieces M, N, which were polished and lubricated with tallow in order to facilitate a circular movement around the stake.

On August 16, with all these preparatory works completed, we proceeded to set up the derricks and capstans. As usual, I was counting on the help of the artillerymen when someone came up to let me know that they could not participate in this maneuver, which required a great number of arms, and more so, expert and intelligent arms. Caught off guard at the moment when the apparatus was ready to work, when all the expenses had been incurred, I called upon the masons who were working on the viaduct. For the first time, these men ended up pushing the bars of a capstan.

At 12:30, the system of forces was put in place. After ten seconds, several spectators who were following the maneuver with great attention pointed out to me a major problem: the fixed points of the derricks gave way under the force of traction; the stakes driven into a moving terrain which heavy rain had weakened no longer held in the soil. It became necessary to create a new anchoring point. I had barely indicated to the chief the installation drawn on Plates VIII, IX, figure 2, when the carpenters, leaving their capstan bars, ran into the ditch with several pieces of wood and the tools necessary to work them. Half an hour later, everything was finished and the obelisk was proceeding toward its destination. It had to be first transported on the cradle and set in such a position that its center of gravity corresponded vertically to the top of the stake, which was done in less than three hours and with great precision.

Having arrived at this point, the workmen had to change the apparatuses and tie the monolith to the turning cradle. When this had been done, two pulley blocks holding two capstans, and whose mobile pulleys were attached to the top and the base of the obelisk,

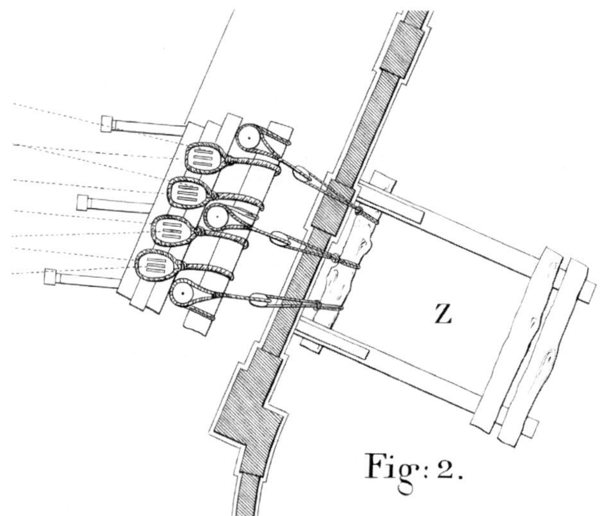

Detail from Plate IX, figure 2: Derricks at a weak point of the rain-soaked ground, which had to be strengthened by carpenters

caused the whole system to turn at a right angle. The circular movement was caused by the impact as the battering rams struck in opposite directions on the extremities of the cradle.

It then became necessary to prepare a new cradle to compensate for the different levels between the cradle carrying the monolith and the new slope, NN. The obelisk was placed on it on September 8.

The Fourth Movement of the Obelisk (Plates VIII, IX, and X)

In these various maneuvers, we had used four capstans and two rams, hitting the back of the cradle. But to bring the obelisk up the viaduct, and raise it on its pedestal, we decided that the motive force would be provided by a steam engine.

No occasion could provide a more brilliant or solemn chance to demonstrate in front of the eyes of the whole assembled population the power of this marvelous agent. It would have been a very imposing spectacle to see a load of 500,000 [pounds] rise majestically in space, without any help from muscle power, to straighten it on its base with the help of one of the most powerful inventions of modern times, the only one, perhaps, that antiquity cannot claim priority over. Unfortunately, the problem that befell the machine while we were testing it, to make the obelisk move, forced us to give up using it.* We had to immediately

* These reflections bring us to cite a few lines from a remarkable article on the erection of the obelisk of Luxor from the pen of Mr. Michel Chevalier, placed in the *Journal des Débats* for October 16, 1836.

"In the beginning, it had been decided that the motive force would be furnished by a steam engine. Unhappily, the boilers of the one that had been installed were found to be insufficient. They did not produce enough steam for the machine to generate all the force of which it was capable. From the experience gained from moving the obelisk toward its pedestal on the inclined plane, it was thought prudent to abandon it. It is unfortunate that sufficient precautions had not been taken in advance by the builder for his machine to generate all its power; it is fair to say that this would have been an easy thing to do. It was a wonderful idea to inaugurate the steam engine on such a solemn occasion. For part of the public, the steam engine is unknown, a kind of mysterious and formidable creation, able to burst like thunder. It was good to associate the monuments of ancient arts with one of the most beautiful products of the inventive mind of modern times. It would have been good to show to 200,000 people one of those machines so wrongly feared by the common people; seizing, without trouble, the obelisk of Sesostris, and lifting it, little by little, without commotion, with a perfect regularity, without the help of any living being, except for a stoker in charge of feeding the coal to the furnace, soul of the machine. Machines are destined to free humankind from all work that demands only brute force—and even, as they are perfected, the work that demands the help of an intelligent being. The steam engine is one of man's most beautiful conquests over nature. This is nature enslaved, working for man and in his place. It is nature made slave, and it is the only slave, the only serf, of the future. The substitution of the steam engine in the place of the arms of three or four hundred men would have taught a complete lesson. It is annoying that people did not foresee a small detail, which forced the people in charge to give it up. Indeed, the partisans of modern mechanics can console themselves for the failure that steam has just encountered, by the triumphs that it continues to experience on land and sea. The modern mechanics will in any case be present in the installation of the obelisk through the ingenious apparatuses that it imagined, instead of the complicated scaffolds that had been used before."

take new measurements to establish other fixed points, etc., etc. We finally had to use the same system of hauling, for which nothing had been prepared. It was important, however, that these preparatory works be done with great haste. The season was advancing; the slightest delay would have forced us to postpone the operations to the following year. So, three days later, the apparatus was ready to function.

Four capstans were installed at the beginning of the viaduct. The head of a derrick was wrapped around each one. The pulleys of these apparatuses were attached to two joists, one of which was attached to the cradle. The other, placed on the pedestal, was held by eight chains, which, leaving from the opposite side of the ramp, were attached to stakes driven into the ground.

At 10:00 a.m., 120 artillerymen, spread out on the capstans, were moving the monument on the inclined plane. In its march upward, this enormous mass was, so to speak, almost floating on the cradle. The force exerted by three men, pushing the base or the top of the obelisk with their shoulders, was sufficient to straighten it, to make it turn right or left, according to need, in such a way as to keep it going in a perfect direction. That was a crucial point. The position of the obelisk on the pedestal depended, obviously, on the direction it would be facing at the top of the ramp.

In less than five hours the obelisk had traversed the length of the viaduct. It had arrived two centimeters from the point we wanted. Until then it had advanced on the inclined plane in forty-centimeter increments of length, which now had to be reduced in this circumstance to two centimeters. To this effect, after having tightened the derricks we stopped the action of the machinery, and with the help of the battering rams we completed this last movement two times. The edge of the base coincided exactly with the side of the square that it was going to occupy on the base, and the axes of the pedestal and those of the monolith were situated in the same vertical plane.

On these two conditions, the success of the last maneuvers, considered as the most difficult and dangerous, necessarily depended. In my opinion, however, the problem was solved. The main difficulty, yet the one about which we worried the least, was to take the obelisk without accident from the ship to the top of the pedestal, and in such a way that we only had to turn it around one of the edges of the base for it to be on its final spot.

Setting Up the Erecting Equipment (Plates XI, XII, and XIII)

In giving an account of the taking down of the obelisk, we described the way in which the wooden pivot had been set (see page 60 and Plate XI, figure 1.) The reverse operation required similar dispositions. Thus, we had to saw off a part of the pedestal, to remove Fragment F so as to place there the pivot and its support.

I had proposed two ways in order to avoid this cutting in the last seating of the pedestal. They were rejected, in preference for a simpler procedure, less

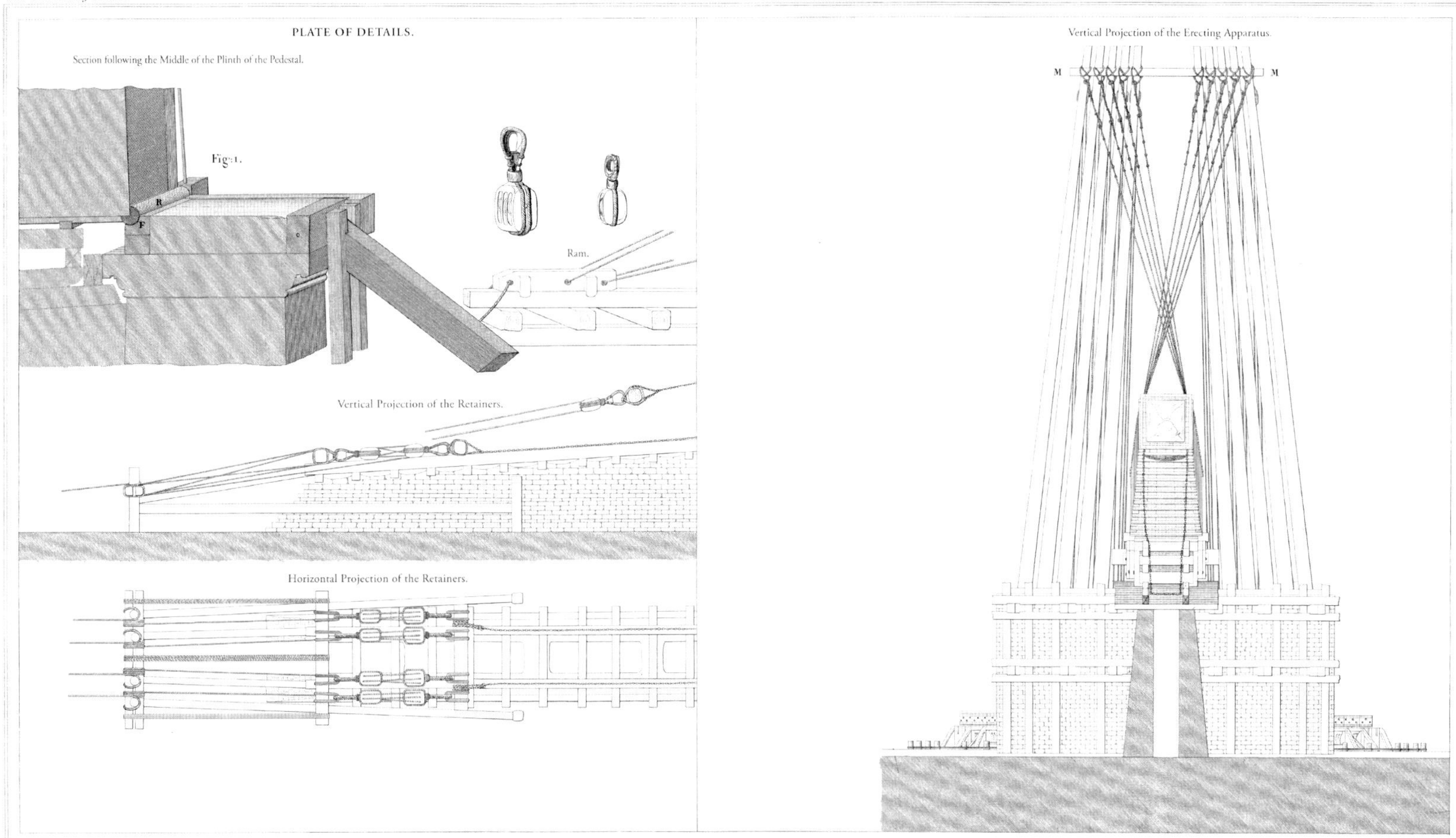

Gravé p. pierrep. Noel. Imp. de Lemercier, Benard et Cie.

Plate XI: Plate of details

Gravé sur pierre par Noël Imp. de Lemercier, Bénard et Cie

Plate XII: Vertical projection of the erecting apparatus

HORIZONTAL PROJECTION OF THE ERECTING APPARATUS

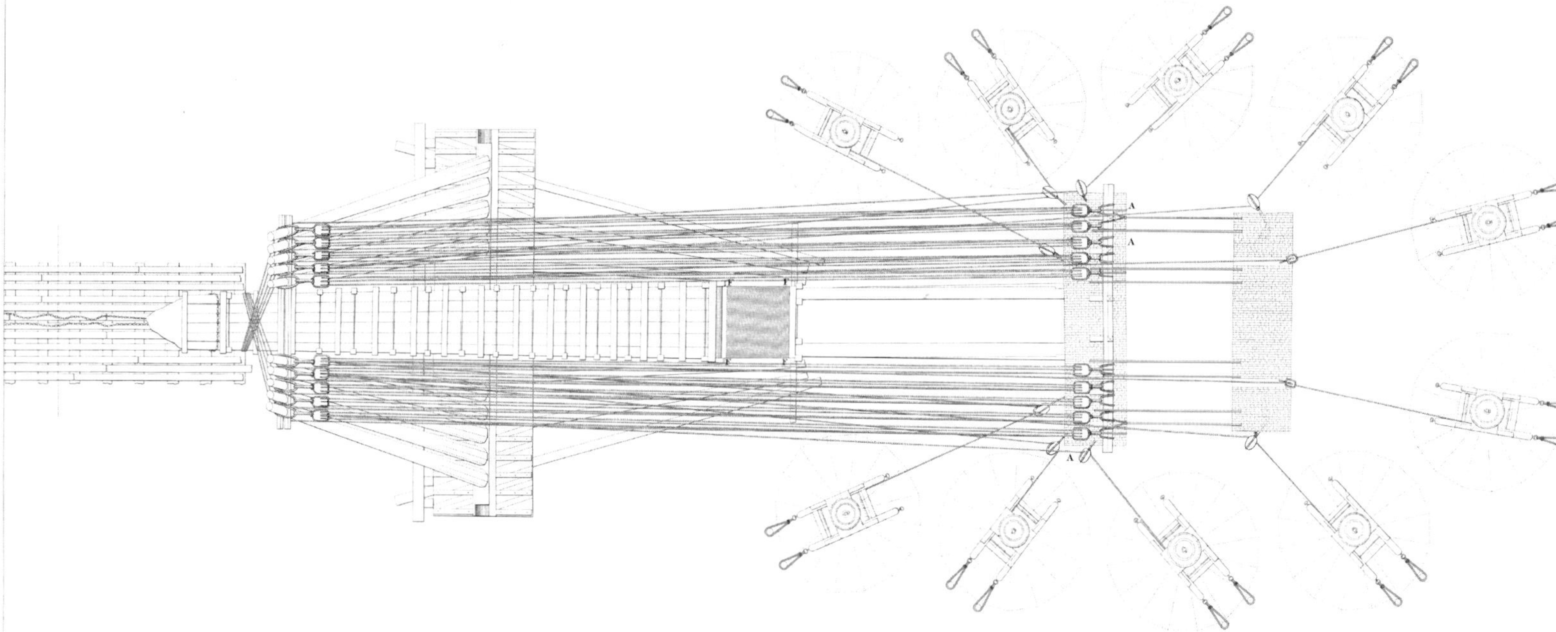

Gravé sur pierre par Noël. Imp. de Lemercier, Benard et Cie

Plate XIII: Horizontal projection of the erecting apparatus

costly, and mainly safer. In the end, I must say that if the monument had not had a crack, the pedestal would not have been cut and one of the two ways would certainly have been adopted.

The pivot being placed as we have shown in Plate XI, we had to make the obelisk turn around roller R to straighten it on its base.

The apparatus responsible for the turning motion was composed of the main parts of the one we used in Luxor. Without returning to those details, we will simply refer to them and give some explanations on the setting specific to raising the monolith.

The obelisk is resting on the sloping cradle, which has been built to take it to the top of the pedestal. A scaffold formed by the union of ten masts or shears, five on one side, five on the other, rests on a wall of rubble, whose axis is perpendicular to that of the viaduct. The base of this scaffold can turn on itself like a kind of pivot. On the top tie-beam MM are attached ten cables, or shrouds, which are wound like a necktie under the monolith, 1.5 meters from the top. Opposite these cables, the tie-beam is attached to ten cables whose mobile pulleys are attached to the top of each mast. These pulley blocks contain seven branches; the heads of the tackle, having passed through a return pulley attached below, close to the ground, wind around the bells of ten capstans arranged in an elliptical space.

When these mechanisms operate, the scaffold, through their action, will lean toward them, raising the head of the obelisk, to which it is firmly tied by the shrouds. The monolith will follow it, turning around its pivot, upon which it will rest during the whole rotation.

Arriving at the position where the height of the center of gravity will go beyond the base, the difficulty will no longer be to raise the monolith. Quite the contrary, it will consist in restraining it with the help of a system that will keep it from going in the direction opposite to the one it has just followed.

This apparatus consisted of four retaining chains winding above the top of the obelisk, like a tie on top of the shrouds. These iron cables grip the roller placed at the end of the inclined plane and submit it to the force of four shears whose fixed pulleys are attached to stakes anchored at the bottom of the ramp.

The lengths of the chains and ropes were calculated in such a way that the pulley blocks are tightened at the precise moment when the mass reaches its equilibrium on the pivot. From this moment on, the heads of the tackle were to be let out, little by little and with great precision, in order to avoid all unwanted acceleration in the movement.

Now we will discuss this plan, set out the theoretical results, list the means of execution, and show the difficulties and the measures to overcome them. To this effect, we will consider the apparatus in equilibrium in one of the positions that it must occupy during rotation.

We first notice that the mass is held in space by a system of ropes that pulls its top in an acute angle. The result is a pressure on the pivot. Part of this pressure is exerted on the socle, as would a horizontal force that would cause this seat to move on the side opposite the mass, which is, with reference to the actual setting, on the side of the Madeleine. In order to prevent this movement, which would have caused serious damage to the monolith as well as to the workers, we braced this block between the two strong beams. These beams are bolted to the cradle, which is itself tied to the fixed cradle by St. Andrew's crosses, with the purpose of rendering the whole system solid. This seat is, in addition, strongly held by two enormous supports. The foot of these struts rests on a fixed point, A. Their top is held by hanging crossbeams that form, with other parts whose ends are engaged under the pedestal, a rigid triangle. The strength of such a system is clear; nevertheless, it barely offered the degree of resistance necessary to bear the force, a force that was measured at more than one hundred thousand kilograms.

If one reflects on the intensity of this force, which had to move a block of small dimensions, merely resting on the ledge, and raised nine meters above the ground, one would easily agree that this was the most important part of the apparatus and the most difficult to strengthen, since it had to be perfectly immobile in every sense of the word. It goes without saying that we could have made, at a smaller cost, a stronger restraining mechanism in the construction of the pedestal itself, but in the situation we were in, this was out of the question.

The combined action of the pulleys and shrouds had to produce on the scaffold a force that tended to push the pivot toward the river. But because this component was tied strongly and indestructibly into the whole system by the beams of fir planks that framed the masonry and pedestal, this part was capable of resisting a much greater force, as it is easy to see by examining Plates XI, XII, and XIII.

Let us look at the calculations of force and resistance. The weight of the obelisk is estimated to be 250,000 kg. For the entire duration of the rotation it will rest on the pedestal; one part of the weight will thus be supported by this resting point. The remaining weight, which makes up the resistance to overcome, obviously depends on its inclination and on the angle at which it will be pulled by the collar of the shrouds. It follows from these given conditions that, to overcome this resistance, it will be necessary to transmit to the shrouds successive tensions represented by 124,000 kg, 121,800 kg, 120,000 kg, 114,000 kg, 100,000 kg, 76,000 kg, and 42,400 kg. At the instant when the height of the center of gravity of the mass will coincide with the points of the space 1, 2, 3, 4, 5, 6, 7, these tensions, by reason of the angles that the direction of ropes form with the plane of the scaffolding, correspond for the same positions to the traction of the apparatuses measured as 104,000 kg, 89,600 kg, 88,000 kg, 86,200 kg, 80,000 kg, 66,000 kg, and 38,000 kg. Thus the force that the mechanisms will have to produce will only be 104,000 kg, and will gradually diminish. Each capstan is moved by 48 artillerymen. Assuming the force of a man to be 12 kg, the ten capstans together will pull the top of the tie-beam of the scaffolding with a force of 115,000 kg, which is 11,000 kg more than the maximum force needed. Since a strong man can exert a force of 15 kg, or even 20 kg, instead of the 12 kg I had assumed, it follows that, if needed, three hundred men would have been sufficient for the operation.

Two fixed points are established behind the pedestal. The first one is composed of two rows of stakes tied between wood beams. The second one, placed several meters farther, is fixed by only one row of stakes, also tied between two beams.

In order to reduce the angle at which the scaffolding is pulled, and thus the tension on the apparatuses, the fixed pulleys of the derricks are placed on a scaffold whose base rests on the first resting point. The eye in the rope of each one of them, the grommet, is pinned to the two platforms, which have been loaded with 104 tons of iron.

It is estimated that each stake, taken by itself, would resist a pull of 3,000 kg. Together, and including the ballast, it is thus capable of withstanding at least a force equal to 194,000 kg, which is greater than the force on those two fixed points of 90,000 kg.

The appendix shows that the thickness of the ropes is more than adequate to withstand the tensions to which they will be subjected during the entire rotation of the monolith. Such are the main results deduced from the theory.

The most important work, and the most difficult, consisted in the raising of the scaffold. Since the transport and erection of such a mass, eight meters above the ground, could only be done with the help of a complicated machine, we set up separately all the elements of which it is composed. The joining of the foot of the shear on the half-cylinder was done by means of two derricks placed one on the ground and the other on the wall itself. The first one would grab the mast and carry its center of gravity to the level wall. Then, the second would raise its socle above the corresponding mortise. It thus became easy to move the tenon to the right and to the left in such a way as to fit it into the cavity prepared ahead of time on the semi-circular pivot.

With the scaffold raised and temporarily anchored, the sailors would proceed to the installation of the ropes. These men, suspended in air at the top of the shears, worked for fifteen days threading the shrouds, setting the moorings, etc., etc.—and, let us not forget, without the slightest accident.

We have now arrived at October 24. The apparatus is finished; there remains nothing but to try it and to agree upon the signals to regulate the working of the capstans.

At noon, 350 artillerymen, their chiefs in the lead, are positioned on the bars of the capstans. Captain Meunier, former student at the École Polytechnique, takes his place at the center of the elliptical space. This officer is in charge of communicating the commands that must be executed at the sound of a trumpet. Several minutes later,

I give the signal [to the men] to walk slowly. Immediately the clarions sound, the capstans turn on their axes, the pulley blocks strain, the tension accumulates, and soon the scaffolding of the shears bends slightly toward the Seine, then straightens itself almost imperceptibly due to the winding of the tackle. This is repeated seven times, and pulls up the top of the monolith. The master carpenter, positioned to watch, cries out: "The obelisk has left!" The monument had, indeed, left its cradle and rested only on the shrouds and on the pivot of the base. Since all was going well, I wanted to continue the maneuver, favored by a beautiful fall day, perhaps the last of the year, but to my great regret, superior orders forced me to postpone the rest of the operation to the following day.

Erection of the Obelisk

From early in the morning of October 25, more than 200,000 spectators poured into the Place de la Concorde, in all the exits, on the terraces of the Tuileries, in the Champs-Élysées. They had been awaiting the erection of the obelisk with avid curiosity. From the time it was announced eight days ago, it seemed that the entire Parisian population wanted to be present for the last act of the drama that began three years earlier among the ruins of Thebes.

This drama might yet fall victim to a terrible mishap. An order poorly misunderstood, a cable badly lashed, a defective piece of wood, a twisted or broken bolt, a friction or resistance poorly calculated, any of a thousand

CLEOPATRA'S NEEDLE — RAISING THE OBELISK OF LUXOR IN THE PLACE DE LA CONCORDE, PARIS, OCTOBER 25, 1836

The obelisk being raised on October 25, 1836 (from *The Graphic*, September 29, 1877)

unexpected accidents could provide the occasion for a dreadful catastrophe: The obelisk broken, millions lost, more than a hundred workmen inevitably crushed by the falling mechanism. These would have been the consequences that could have followed the failure of this operation. It would have been enough, without a doubt, to disturb the firmest spirit; and despite the security that the means of execution gave me, I admit I was not without anxiety, thinking of the grave responsibility that rested upon me.

The weather was dark, but with no threat of rain. This was an important point.

Before proceeding with the grand maneuver, we placed in a cavity hollowed out in the center of the base a cedar box containing current gold and silver coins, and two medals with the likeness of the king and bearing this inscription: "Under the reign of Louis Philippe I, King of the French, M. de Gasparin being Minister of the Interior, the obelisk of Luxor was raised on its pedestal on October 25, 1836 by the efforts of M. Apollinaire Lebas, marine engineer."

At 11:30, the artillery sounded their bugles and began their rhythmic circular marching; then the tip of the needle left the cradle, rose progressively, and described a great ascendant arc, while the support, which had been vertical, inclined little by little in the direction of the force and described an arc opposite to that of the obelisk. The pivot on the base turned on itself almost imperceptibly, causing a great spurt of tallow and wood sap out of the cracks, so great was the compression that it underwent in its casing.

At noon the king, queen, and royal family appeared at the Ministry of the Navy and situated themselves on the balcony that had been richly decorated and prepared to receive them. Chants of "Long live . . ." saluted the arrival of Their Majesties.

During this interval, the monolith had gone through an arc of approximately thirty-eight degrees. It was almost near the point when the tension exerted on the hinge, which had gradually increased with the inclination of the monolith, would reach its maximum. The inclination of the monolith began to lessen because of the arc inscribed by the center of gravity. Positioned on the base, I could follow all the operations with a

The medal that was struck to commemorate the raising of the obelisk; courtesy William Joy

glance. For several seconds I felt a slight movement that I attributed at first to an illusion caused by the displacement of the surrounding objects, but at this precise instant, the vibratory movement became more and more pronounced, giving me the certainty that it was produced by the shaking of the block on which I was placed. This discovery was decidedly reassuring, until a cracking sound caused by the compression of the wood could be heard. Immediately I gave the signal to stop in order to search for the cause of this noise and to examine all the parts of the fulcrum one by one. "Nothing has moved," cried Mr. Lepage, inspector of works, "you can continue." Everything was in good order. It was just that the tensions of the two tie-beams F,F were so great that they resonated a little like the strings of a violin. The dovetail *a* was inset by five millimeters on the cradle's transverse by the force, the bolts began to twist, and finally the compression manifested itself on all the surfaces in contact, so much so that the grained wood penetrated three millimeters into the other hard wood. I couldn't believe it. In this state, if the fixed point had given out, or rather if the links that held the base in a fixed position had broken under the action of the force, the obelisk, the pivot, half of the pedestal, and the whole system would have been flung violently toward the Madeleine. The imagination can readily conceive the horrible catastrophe that would have resulted for all the workers taking part in the operation. Be that as it may, despite that warning, despite that sudden stop and unexpected maneuver, at that moment when all the forces on the equipment were in play, no apprehension seemed awakened among the workers. They seemed to be oblivious to their individual concerns. "The timbers secured themselves," said the carpenters, that is all. Order, harmony, and the utmost sustained attention continued to preside over the operation, and no one dreamed for a single instant of leaving his post. I immediately resumed mine on the base and the maneuver resumed.

In order to quickly pass over this slightly unsettling episode that I called the weak point of the system, I gave the signal to move more quickly. The obelisk, continuing its movement, rose majestically without shaking, without noise, like an ordinary load, and traveled another third of its course in less than forty minutes. It was close to the position where it would hang naturally to rest itself on its base. The action of the capstans was thus suspended to prepare for the last phase of the operation.

As early as the day before, I had given orders to rework the temporary cables that restrained the chains near the pyramidion. The late hour did not permit this detail to be carried out, and the order was completely forgotten. This omission urgently needed to be fixed before the equipment could be used. If this rope, which in any case was out of service, had broken under the action of the chains, which it was shifting from their natural direction, its breaking could have caused a collapse, a shaking of the system, and consequently would have compromised the success of the very delicate maneuver. As soon as this problem was detected, two sailors clambered up to the top of the obelisk to dismantle the bridle.

In describing the retaining mechanism, we said that the length of the chains had been calculated so that the ropes were sufficiently tightened at the moment of verticality of the center of gravity of the mass when it would go past the hinge. Theoretically, in this position, the ridge of the base parallel to the swivel would rise 0.5 meters above the top plane of the pedestal, and from this instant the obelisk, which until now had been supported by the shroud wires, would be held up by the iron cables. This change had to happen very slowly, centimeter by centimeter, in order to avoid a crash, a collapse that would have caused serious accidents. In anticipation of this, I asked the specialist carpenter in charge to measure what would be the precise height of the ridge on the base below the stationary base bed. The workman converted this distance, given in feet, to metric, but was mistaken in the calculation and announced 1.10 instead of 0.75 meters, which was the real distance. Assuming that the first number was correct, it was necessary to disengage the pivot from the support, whose angle of rotation corresponded only to a height of 0.95 meters. We immediately began the work. Ten mortise axes wielded by strong arms created a horizontal cut in the hollow piece, when we realized the error that had given rise to this painful and laborious work.

Everything was set for this important operation. The working of the capstans was regulated as follows: First we asked for three turns, then two, then one, then a half, etc., etc., up to the moment where the point of equilibrium was passed, when we saw the chains tighten and the pulley begin to strain. Then it was loosened, little by little and with great restraint; the main pulleys released the cables by degrees, and the obelisk, which five years prior had come down from the base where it had rested for thirty-three centuries, came slowly to rest and without shaking on the rock from Brittany that formed its new pedestal.*

The whole operation took three and a half hours, or rather two and a half hours, deducting the time lost to unnecessary work.

No accident occurred, no one was injured, all happened according to our expectations, according to our provisions and conclusions from our theory. We had, by the way, strongly opposed the proposal to barricade the Place de la Concorde, to keep the public out. This unpopular measure would have reminded people of the *bando* [decree] by the pope on the erection of the Vatican obelisk.† That said, the authorities were not happy to have yielded to our requests, but the day was not marred by any disorder. More than 150,000 people

* The retaining ropes were manned by sailors in the service of the king, under the orders of Mr. Rolland, lieutenant of the marine.

† See part 4, "Fontana."

stood for four hours in spite of their anxiety, with no one showing the slightest sign of impatience.

Throughout the whole time, the site offered the beautiful spectacle of three special units, obeying a single voice, lending each other mutual support, forming a single unity to cooperate for the success of the work. You cannot imagine the joy, the satisfaction, of all the workers, the enthusiasm that they exhibited at the moment when the crowd gave them their well-deserved applause, and of which His Majesty had given the first signal. Permit me to express again my personal gratitude to the skilled subordinates, the carpenters who perfectly executed a new work, difficult, perilous, and who marvelously supported all my intentions; the sailors whose zeal, agility, courage, and capability the Parisian public so well appreciated; the artillery who lent their intelligence to a precise maneuver of which few men are capable. Everyone performed his task with admirable zeal, and for myself, I would like to state here, it is to their heartfelt and complete devotion that I owe the most beautiful day of my life.

While I was walking across the Place de la Concorde to go to the Ministry of the Navy, where the king had summoned me, four of our courageous men had already climbed to the top of the obelisk and raised their tricolor flags and laurel branches. Thousands of witnesses present at this scene know with what electric enthusiasm this signal was welcomed by the crowd—especially when the venerated chief of the great nation, Louis Philippe, removed his cap to salute these glorious colors which, thirty-seven years before, the French army had displayed while applauding at the Theban ruins.

After the king's dinner, to which I had the honor of being invited, His Majesty gave me the sum of 3,000 francs for the workers who had collaborated for the success of the maneuver. From the Château des Tuileries one could see the base and pyramidion of the obelisk, which the carpenters had illuminated.

It is my duty to name here the people who eagerly assisted me in an always difficult task, that of doing something out of the ordinary and in the presence of an immense crowd.

I will first cite Mr. Lepage, inspector of public works. Generous with his efforts for the good of the service, possessing extensive practical and theoretical knowledge, this devoted agent, modest and industrious, greatly contributed to the success of an enterprise that he had taken to heart with an enthusiasm worthy of great praise. His zeal never diminished, and at the moment of greatest danger, he showed the greatest intelligence and calm.

Mr. Heurteloup, the second inspector, only cooperated in the last part of the work, in which the obelisk was the object. I have nothing but praise for his zeal and his activity. Charged with taking precise measurements to determine with exactness the precise position of the obelisk with respect to the axis of the pedestal, he gave me only good results. The plans relating to the obelisk were drawn by this architect with meticulous care. The smallest details were clearly indicated there, the maneuvers perfectly described.

Mr. Pouillet, framework specialist, provided us with excellent materials and very intelligent workers. This contractor did not recoil in the face of difficulties of execution—and let us hasten to say that it is not easy to build a gradient 120 meters long that had to be laid out, adjusted, and placed with precision as perfectly as possible.

I did not lack for any kind of help from the Ministry of the Interior, thanks to the enlightened and altruistic zeal of Mr. Denoue, former student at the École Polytechnique, today the head of the Bureau of Public Works. Possessing perfect understanding, a quick mind, and an education that allowed him to grasp difficulties, this public servant perfectly understood that a work of this type could not be subjected to overwhelming administrative paperwork. Thus, to cite only one point, even before he received the report about the malfunction of the steam engine, Mr. Denoue had hurried to fix it. The same day, a telegram ordered the harbormaster of Le Havre to send to Paris the five capstans that had been stored at this port. I also am indebted to him for providing soldiers to move them. He also had the fortunate idea to get the army artillery—in which he served as an officer—involved in helping.

I wished to pay here my debt of gratitude. I will therefore refrain from mentioning the actions of some jealous souls, few in number, who tried to hinder our work. At most, these annoyances aroused in the public, as they did in me, a smile of pity. I prefer not to lift this veil, and to speak instead of the awards that were accorded to my collaborators.

On November 11, I received the following letter.

> Sir,
>
> The erection of the obelisk of Luxor has obtained the unanimous acclaim of the king and the public. It is with great satisfaction that I send you my congratulations on the success of this important operation.
>
> A medal having been coined to serve as a souvenir, I send you two proofs, one in silver, the other in bronze.
>
> I have the honor to announce to you, at the same time, that you will be given a pension for the care that you gave to the direction of the works, the sum of 4,000 francs.

I have not forgotten, sir, the favorable account that you gave to me of the people who were responsible, under your orders, for the oversight and execution of the details. Rewards proportional to their services will be allotted:

To Mr. Lepage, Second Inspector, 1,500 francs
To Mr. Heurteloup, First Inspector, 1,000
To Mr. Labrie, Foreman of Carpenters, 300
To Mr. Dacheux, Foreman of Sailors, 300
To Mr. Card, Magazine Guard, 300
To Mr. Morel, Foreman of Sailors, 200
To Mr. Mosqueron, as above, 200
To Mr. Monot, Foreman of Carpenters, 200

Please let those mentioned above know of the allocations of which they are recipients, and tell them they will soon be given to them.

Receive, etc.,

Signed, Gasparin

To these honors, I should add the one that Mr. Lepage received after the completion of the work. He was decorated with the cross of the Légion d'Honneur as a reward for his services and good management.

Part 4 — Fontana

"Water on the ropes!" Evaluation of the tradition of this phrase — Examination of Fontana's work — Details of the erection of the obelisk of Rome, and a description of Fontana's Castle

The erection of the Luxor obelisk naturally brought us back to similar events that happened in the past. Thus, several writers decided to compare it to the operation that was executed in Rome in 1586 by Domenico Fontana. They even published several excerpts from the work of this skillful architect, but these extracts do not agree on all the main points of the problem: according to some the length of the obelisk was 111 palms, according to others 113½, 115, etc., etc.

The same difference is noted in the estimate of the weight of the obelisk, and in the number of men assigned to working the capstans, which varies from six hundred to nine hundred men. Finally, some confuse the lowering of the obelisk with its erection. They hold the second operation responsible for the damage that occurred during the first. And remarkably, almost all agree that there was a moment when Fontana almost failed in the project. This circumstance gave rise to a more than doubtful event which subsequent generations accepted as an established and authentic fact. The obelisk, they say, had reached a certain height, as the ropes were stretching due to the enormous weight that they had to support, when a workman within the crowd cried out, against the order of Sixtus V, who had commanded complete silence on pain of the most severe retribution, "Wet the ropes!" It was even said that, to reward him for the fortunate inspiration and devotion, the pope gave this worker the exclusive and hereditary privilege of selling palm fronds on the day that this holiday is celebrated (Palm Sunday), a privilege that it is said his descendants still possess.

I certainly feel all that is grand, admirable, and poetic in this sudden inspiration of a man who divines a law of physics in a strong voice, powerful and courageous, who risked death for the success of a great project, to preserve a precious monument for his country. I would be the first to draw a veil over this anecdote, to regard it as an authentic fact, although it is in opposition to theory and practice, if it did not bring criticism to the reputation of Fontana. This famous *aqua* ("Water!"), which had become known to all, would prove, without objection, that this director would never have realized the goal through his own vision, and that the success of his operation was due solely to a happy chance. However, we should say it is an ingenious story, a poetic fiction, that is easier to believe because it speaks to the soul; it warms it, penetrates it, and inspires a strong emotion that keeps us from thinking.

Every time that I have wanted to fight this historically suspicious and mathematically absurd story, to destroy it, I was answered with poetic assertions to which I could only respond with the cold reasoning of theory and the results of experience. This manner of discourse earned me, and will continue to earn me, the scornful qualification of a sorry calculating machine, a man too positive. I will get over it. I naturally had to deal as a mechanic with a question that belongs to the realm of the mechanic.

The chief tackle block intended to raise a load is provided with a capstan. If one uses this piece of equipment, the ropes lengthen till they are sufficiently tense to raise the mass attached to the mobile pulley. The lengthening and tension are two components, of which one is always the consequence of the other. Having established this, and the weight in question being invariable, whatever its elevation above ground, the ropes will never be subjected to a force that is not the same as the one at the beginning of the movement. Further stretching, such as appears in the story of Fontana, would therefore be an effect without a cause. In reality, the ropes can momentarily resist the pull whose action, continued over a certain time, would end up breaking them. Thus, if you can lift a mass a few centimeters above the ground using derricks or pulley blocks, that doesn't always mean it can be raised higher. If the ropes had been too stretched from the beginning, if one had gone beyond the limit of the elastic force of the hemp, then its physical properties would soon be destroyed and the fibers of which it is made would tear successively. But in this case, all the water of the Seine would not return them to their original state. I am also saying that it would be really absurd to wet the ropes. The result would be the opposite to the one you wanted to achieve. The shrinkage caused by the liquid would increase the tension, already too great, and only hasten the breaking point. The effects produced by the changing hygrometric qualities of the hemp are exaggerated: they are not generally observed except on ropes not under tension, whose length can vary by several feet. But if they are first submitted to strong tension, as were those of Fontana, all their parts will only form a mass almost impenetrable by water. The liquid will only act on the outside threads; the sole effect of their vertical shortening will be to destroy the harmony of the whole without changing the total length of the ropes.

The other objections offered to these demonstrable reasons are not worth refuting seriously.

Convinced, however, that I had to be wrong, against the opinions of everyone, I decided to read the work of Fontana as soon as I completed the work of which I was in charge. Mr. Villain, one of the most distinguished architects of the capital, had the kindness to put this book at my disposal. I got to work. I translated the text literally word for word, starting with the Notice to Readers, in order not to let anything escape me. At each page that I turned I expected to find the explanation of the anecdote that had been so often told to me. My expectation was defeated. I arrived at the end of the book without discovering the least bit of information that had anything to do with this story. And one should not believe that Fontana had tried to obliterate the knowledge of an authentic fact that supposedly happened in front of a whole crowd of people. The simplicity, the naive honesty with which he recounted the difficulties that had occurred during the operation, leaves no doubt about the veracity of his report. This architect doesn't spare us so much as a twisted bolt or broken cotter pin. Everything is told with precise care.

The specific plans are appended to the text. These drawings give us quite a precise idea of the details of the construction. The timbers are worked, bolted, tied together according to the same rules followed today. The pulleys, capstans, etc., are exactly the same instruments we use in maneuvers requiring strength. Finally, Fontana had perfectly appreciated all the difficulties; he was going in the right direction. It is useless to object that the artist had erected on Vatican Square a gigantic castle whose principal columns were more than a meter in diameter, a true forest descended from the Apennines; that the tackle or head of the derrick, passing through three or sometimes four return pulleys, would result in a loss of force equal to about half the effort; that to raise the monolith, he used forty capstans moved by 900 men and 140 horses; that the adjusting of the base on the pedestal required seven days of work, of trial and error, etc. All these objections, although right and perfectly founded, do not weaken in any way the opinion we expressed on the talents of Fontana. If one wants to be fair and impartial, one has to go back to a time when theory and mechanical arts were in their infancy, when science had not yet provided the means to calculate the friction and the stiffness of ropes, when the practitioner could not find inspiration in the past, and could not benefit from studying the experieces, even the mistakes, of his predecessors.

Such was the position of Fontana, and nevertheless, even if it doesn't please lovers of fairy tales, this skillful architect, in order to succeed, and succeed completely, did not need the miraculous help of a few drops of water. To be truthful, I had guessed that for a long time, but today my conviction on that point is completely established.

I had started the translation of Fontana's work only in order to enlighten a historical fact. Later on, I thought it might interest the public for the link to arts and culture of another time. I also had the ambition to reestablish Fontana in all his glory. These considerations made me present a short but faithful summary of the memoire published by this architect in 1590, titled *Della transportatione dell' obelisco Vaticano et delle fabriche di nostro signore papa Sisto V, fatte dal cavaliere Domenico Fontana architecto di sua santita.*

I added to it the explanatory drawings necessary for the understanding of the text.

In his address to the reader, Fontana sets out the reasons that led the pope to have the Vatican obelisk transported to one of the squares of Rome. Sixtus V, he said, the best and wisest of princes, was barely on the throne when he resolved to destroy and erase all traces of idolatry, to purify the obelisks, the colossi, and all the superb monuments raised by the pagans in honor of their gods. Animated by an ardent zeal for the holy religion, our Holy Father decided to start the holy enterprise on the obelisk of St. Peter. He wanted to purge the needle from the stain of infamy, this needle consecrated to pagan superstition, by making it serve as a support for the Holy Cross, a simulacrum abhorred by the gentiles as the instrument of the most terrible torture, and become, through the death of Christ, a triumphant symbol of kings and emperors. In sum, he wanted to transform the column of Sesostris into a Christian monument, to make it a trophy of Jesus Christ.

There is no profane thought in this presentation. The moving of the obelisk is motivated solely by religious ideas. There is not a word that refers to art, science, or the convenience of placing the obelisk in front of such and such an edifice. In our days, we discussed the place where it would be convenient to erect the obelisk of Luxor in Paris, but the points on which this discussion rested are completely foreign to those that were the object of the previous discourse. This comparison seems rather strange to me.

After this foreword, Fontana goes into several explanations regarding the plan of his work. He undertook it, he says, only to fill in the silence of antiquity, to leave several documents useful to engineers who would be charged with moving heavy loads.

We find in other chapters the following details.

Of all the obelisks that Rome had taken from Egypt, only one had remained standing and whole among the ruins of its antique magnificence. It stood in the former circus of Nero, on a base buried in debris, its axis leaning two palms toward the nave of St. Peters. This square was muddy, little used, and so far from the center of traffic that most of the curious who came to the capital to admire its monuments would return home without having seen this marvel. The idea of moving this monolith had existed from the very early time of

Christianity. But the first architects and engineers of our era—*pocco valenti* (little skilled), says Fontana—lacked information, especially about the processes followed by the ancients. They faltered in the face of the difficulties of an enterprise whose execution could only be the result of a miracle. Later the popes, in particular Nicolas V, Paul II, Paul III, and Gregory XIII, embraced this idea, but when these same obstacles presented themselves, they could find no one who would agree to undertake this extraordinary work and assume such a great responsibility. Sixtus V, more daring, more tenacious than his predecessors, convinced that the moment had come to accomplish this gigantic task that would immortalize his pontificate, decided to put it to execution. A commission composed of the most distinguished savants, architects, engineers, and practitioners was given the task of determining where to put the obelisk, and more particularly, to come up with a plan for the move that would be immune to all objections. This commission convened on August 24, 1585. Since none of the members had presented a general plan with details upon which a discussion could start, they suggested several ideas; they spoke, they discussed at length, everybody tried to impress the others by his erudition, and they ended without any conclusions. Sixtus V did not allow himself to be deterred by this first setback. He called upon all the talents and luminaries of the century. More than five hundred artists from different countries attended the second meeting, which took place on September 25. Some came from Milan, Venice, Florence, and others from Lucca, Como, Sicily, and even Rhodes and Greece. Each presented a plan, a model, or an essay.

Despite the differences of opinions put forward to solve such a significant problem, most of the competitors agreed on this point: that it was more certain, easier, more prudent, to move the obelisk standing up rather than to take it down and raise it up on a new base. A few proposed to transport it, not only standing, but with its base.

Others had adopted a compromise solution. They wanted to move the monolith neither standing nor lying down, but at a forty-five-degree angle. Those who proposed this middle solution proposed to lift it with only one lever, serving as a balance beam, others with the help of a screw or wheel, etc., etc. It just goes to show that in all times, in all countries, the erection of an obelisk will give rise to a plethora of absurd ideas.

Fontana was among the competitors. He used a model in wood with a lead obelisk, reduced to proportional scale, to perform in front of this solemn assembly the two operations to be performed on the obelisk. His opinion was that it had to be laid horizontally, then dragged on rollers to the top of the pedestal, then erected on its base using derricks and finally capstans. This opinion, at first strongly opposed, won. The commission decided unanimously that the method proposed by Fontana was preferable to all the others.

However, a few members, a little concerned about the difficulties that had occurred to them during the discussions, thought that Fontana was still too young to command such a complicated engineering project.* They proposed to give the project to two architects who were skilled in the art of moving heavy loads; they would be expected to consult with the author of the project, in order to devise the method of execution. This proposal was approved and Fontana went home delighted to have won the prize. However, his joy and satisfaction didn't last long. He tells us that, starting the next day, after considerable thought, he postponed going before the Holy Father to start the work for seven days. Taking advantage of the pretext provided by his two colleagues, he accompanied them to Monte Cavallo.

When the pontiff questioned him about the obelisk, he answered that he found it impossible to think about a problem that demanded all his attention. "At this moment," he added, "only one idea preoccupies me, absorbs my intellectual capacities. I fear that they will bring some modification to my plan that could cause serious accidents, for which I would be partly responsible. In truth, the more I think about it, the more I am convinced that they did me an injustice, because no one can execute an invention better than its inventor himself."

Sixtus V, struck by the correctness of his observations, ordered that he would be the only one in charge of this operation. Fontana immediately went to Vatican Square. He was accompanied by twenty-five workers who started to dig the foundations of the new pedestal.

They threw into these foundations several commemorative medals. Two travertine caskets, each one containing twelve medals, were dedicated at this inauguration. Each medal had on one side the image of Our Lord, on the other different symbols. The following subjects were represented:

A man asleep under a tree with the inscription: *Perfecta securitas* [perfect safety].

Three mountains, each with something on top. The one on the right, surmounted by a cornucopia; the one on the left by a laurel branch; the middle one by a sword whose point is turned toward the heavens and serves as a support for a scale. The legend at the bottom was: *Fecit in monte convivium pinguium* [He has made a feast of fat on the mountain].

St. Francis kneeling in front of a ruined church, carrying a crucifix in his hand with the legend: *Vade Francise, et me repara* [Go, Francis, and repair me].

Finally, on the faces of a few of them was the portrait of Sixtus V, and on the reverse, Religion and Justice.

Impatient to finally see resolved the difficulties that, deep down, he had never considered as insurmountable, Sixtus V gave to Fontana a brief by which he authorized him to demolish all the houses that would be in the way of the work, and to obtain in Rome, and in the towns of the Holy See, the materials and instruments of all kinds,

* Fontana was then forty-two years old.

food, etc., that he thought he needed, for which the owners would later be compensated. This brief relieved him of the salt tax and of the administrative formalities that always caused delays. He strongly advised all the workers, without exception, on pain of suffering the disgrace of the pontiff, to help, to assist the architect, and to obey him should he need their assistance. Finally, he gave Fontana the right to anything he might judge necessary to ensure the success of his enterprise and put at his disposal the resources of the government.

With these powers, Fontana immediately sent devoted and capable agents to different places. They were charged to prepare and send to the capital, as soon as possible, all the materials for the execution of this project. From the Campo Morto forest, located twenty miles from Rome, he got enormous beams that were transported on carts with huge wheels pulled by seven pairs of oxen.

Fontana wanted to supervise personally the execution of all the parts of his apparatus, and particularly the making of the ropes that had to play such an important role. He planned so well that the work was started all at the same time. All was arranged, set, and coordinated in such a way that, without any delay, all the preparations were completed on the same day.

Before entering into the details of the construction, we must set out the basics of the problem.

We have said that the needle rested on four iron castings attached to the top of the pedestal, which was completely buried in rubble.

Here are the main dimensions of the monolith taken from the work of Fontana.

Length measured from the base up to the beginning of the pyramidion	107 ½ palms
Width of the side at the bottom	12 1/12 palms
Width of the side at the top	8 1/12 palms
Height of the pyramidion	6 palms
Total height	113 ⅓ palms

From these dimensions it can be deduced that the volume of the obelisk is equal to 123 cubic meters. This gives a total weight of 333,100 kilograms, by taking the estimate of 2.70 for the density of Syene granite, as we did before for the monolith of Luxor.

Comparing the height of the pyramidion with the dimensions of its base, Fontana concluded, with Pliny, that it was partly broken during the erection of the monument. To confirm this assertion, he cites the measurements taken on the other monoliths. In all of them, the height of the pyramidion is equal to 1.5 times the side of the upper square. The needle of St. Peter is the only one that strays from this rule of proportion. Secondly, it is plain to see that this part of the obelisk was not carved by the same worker. Its faces were rough and, so to speak crude, whereas those of the shaft have been polished perfectly.*

* The obelisk of Luxor presents the same defects in the dimensions of the pyramidion and in the execution of its faces.

After having calculated the weight of the obelisk, Fontana believed he had to do a few trials in order to determine exactly the power of each capstan. A second purpose of these experiments was to coordinate, to set up operations in such a way that the maximum force transmitted to the rope by the action of the men and horses was always less than the elastic force of the hemp. By this wise foresight, it became impossible for the top or tackle of the derrick to break during the maneuver. This is what several of the engineers did not understand, and they made a thousand objections on this point. Convinced that it was impossible to have several machines act simultaneously, to give them a uniform speed, they did not even bother to examine or analyze the precautions taken to prevent this problem, and persisted in this error where this stubborn idea had pushed them. One can prove anything that one wants with words, and nothing is more flexible in this respect than theory. But, armed with the fact that he experimented, a good architect does not let himself be influenced by opinions, which are often inspired by jealousy. He continues his work. This is what Fontana did.

On the basis of positive data, given by experience, this architect calculated that using derricks, each with forty capstans moved by 970 men and 75 horses, using five levers, each handled by fifty-three men, and with wedges inserted under the base of the monolith, enough force should be produced to lift it. "If, against our expectations," he added "the moving of this mass required a greater effort, one could increase the number of machines." Before lowering the obelisk, Fontana intended to lift it vertically three palms above the pedestal in order to place under its base a platform on rollers onto which it would be lowered.

Committed on this first point, a support to engage the forty fixed pulleys of the derricks had to be created in this space. This called for the construction of a huge scaffold, which was called Fontana's Castle.

The details into which we are getting can only be properly understood with the aid of the plans drawn on Plates XIV and XV. The last one represents the horizontal view of St. Peter's Square from the level of the pedestal. The two others give an idea of the castle that was used both to lower and to raise the obelisk.

Description of the Castle

In order to avoid settling, two rows of beams on top of each other in the shape of a cross were placed between the extended sides of the lower square of the pedestal. The four platforms built like this had to support the entire apparatus.

The castle, not including the roof, was 123 palms high above the foundations. The main part of this large scaffold consisted of eight columns or antennae erected vertically in front of the obelisk, four on one side, four on the other. In order for these columns to stay perfectly straight, it was necessary to support them with forty-eight braces or struts, tied together and also to the

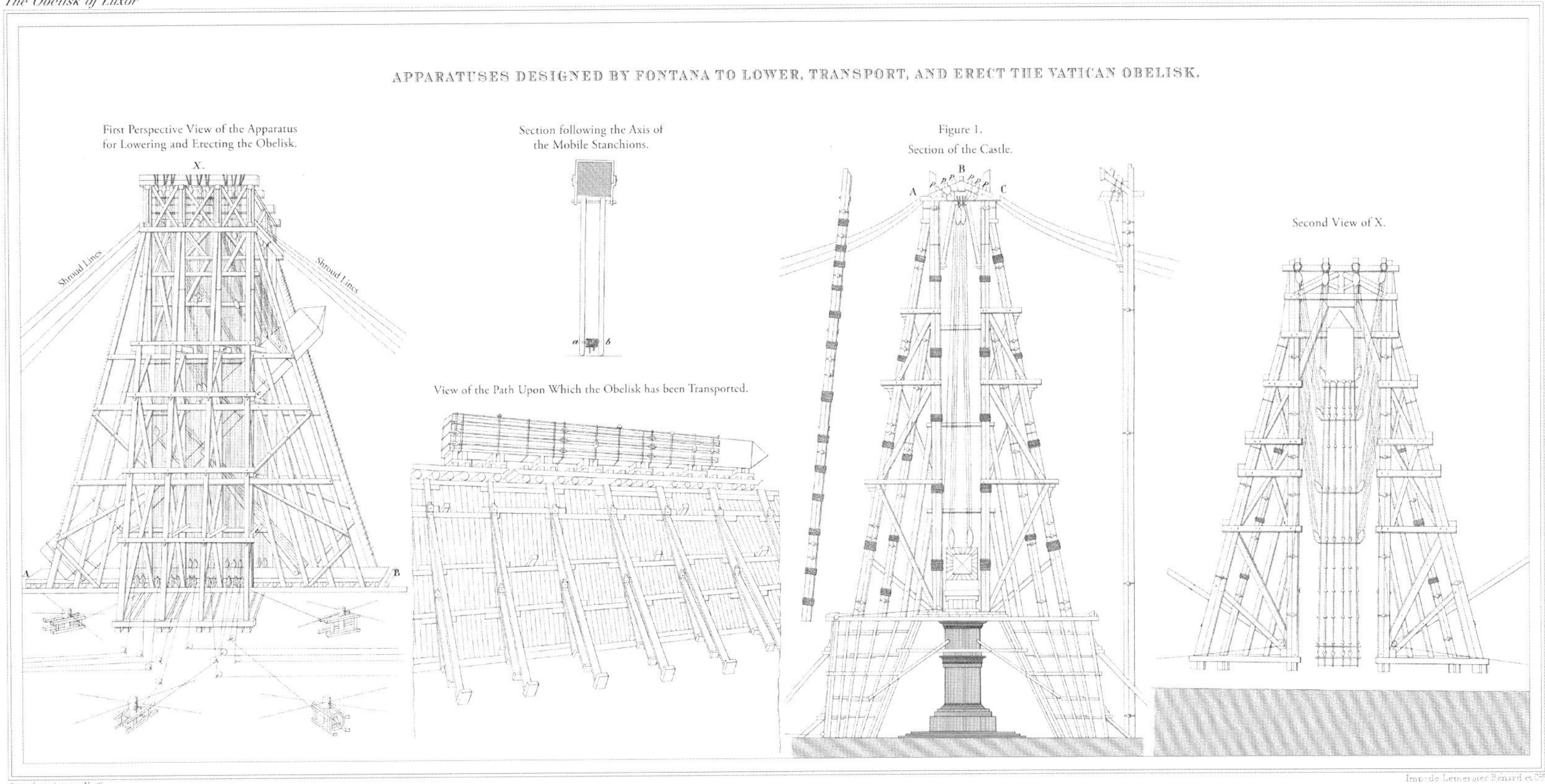

Plate XIV: Apparatuses designed by Fontana to lower, transport, and erect the Vatican obelisk

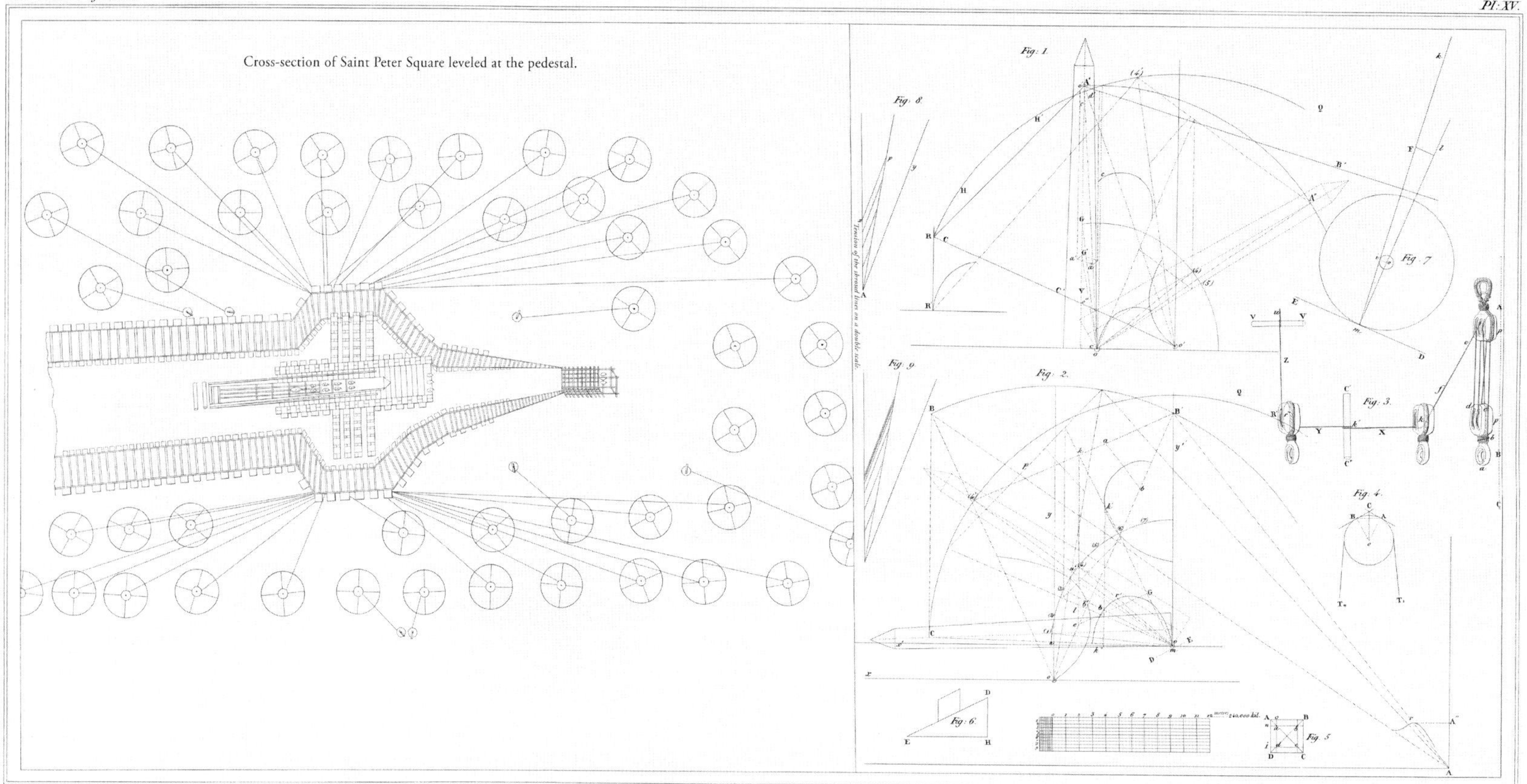

Gravé sur pierre par Noel.

Imp: de Lemercier, Bénard et Cie

Plate XV: Cross-section of Saint Peter Square leveled at the pedestal

antennae by St. Andrew's crosses, crossbeams, iron collars, lashings, etc., etc.

Each column, because of its dimensions, was composed of several pieces of wood. This was a real assembling derrick, formed in its thickness by four beams placed flat, without tenons and mortices, whose transverse section was a square, each side measuring 4½ palms. These beams, placed end to end, were tied in a four-sided bundle by means of a linchpin, iron, and rope collars, distributed along the length. This whole system was tied tightly, by inserting wedges in the voids between the derrick and the ties just mentioned. These details of execution are shown in Plate XV, figure 1. A similar process for the construction of the first and second rows of stations was followed, for the same reason.

Eight trusses ABC bound the tops of the antennae. They were supported by six horizontal beams *p,p,p* more than three palms wide. The fixed pulleys of the derricks were attached to these pieces of wood.

The antennae, attached in this way, could not bend on the outside at all. But pushed from the outside to the inside at the points corresponding to the tip of the brackets, they had a tendency to bend in the opposite direction. In order to avoid this problem, Fontana had this part strengthened. And how? By means of a cross-piece nailed to the column and buttressed by the horizontal braces *e,e*. Finally, by an excess of precaution justified only by the importance of the operation, they added eight large rope shrouds at the extremity of the castle.

When the scaffold was finished, Fontana, delighted with his work, compared its strength to that of a boulder or a massif, built ad hoc to support a great construction.

While the castle was rising in space, the area was being leveled. They took down the houses that were in the way of the installation of the capstans, and they were also working on the cladding of the obelisk. Already its shaft had disappeared beneath a double row of mats and thick planks. Iron bars, crisscrossing each other at right angles under the base of the monolith, encircled it and came up against each of the faces of the cladding. At intervals along the bars were hooks on which abutted nine iron bands. These bands, while they were intended to tighten the armature against the granite, were also to serve as support points for the forty mobile derricks. With these arrangements, the whole system was stable and immovably tied to the mass. This condition was necessary, as we will see later.

The preparations for the lowering were completed on April 28, 1585.

The public curiosity, deeply excited by such an important operation, had attracted to Rome a great number of foreigners. The public authorities felt obligated to take precautions in order to avoid disturbances due to a huge multitude assembled in one place. They had all the streets ending on the square barricaded. A *bando* of the pope, published two days before the raising, provided for the death penalty on anyone who crossed the barrier to the work site. The most severe punishments were levied on those who would impede the workers in any way. It was also forbidden to talk, to converse, to make noise, in other words, to disturb the silence. The head of the police had received the order to station his colleagues in one of the corners of the enclosure to deal severely with offenders.

On April 30, two hours before daylight, two masses were celebrated to implore the enlightenment of the Holy Spirit. Fontana, together with all his workers, took communion. The day before, he had received the benediction of the Holy Father.

Before dawn, the workers had assembled on the worksite. Fontana exhorted them to do their duty, to execute with precision and intelligence the commands that they would receive, and reminded them of the agreed-upon signals.

At the first sound of the trumpet, all the capstans must turn together; the signal to stop would be announced by the tolling of the castle's bell. After this short speech, he wanted to distribute the workers to their posts himself, to examine for the last time all the parts of the apparatus, which, as discussed above, had to lift the obelisk three palms above the pedestal.

Among the measures taken to ensure the success of this maneuver, we will only mention that a group of carpenters was responsible for constantly beating wood and iron wedges under the base of the monolith. "Everybody knows," says Fontana, "the power of wedges. I used them for the purpose of helping the apparatus, by the action of an active force, to create four points of support, which permitted us to lift the mass without suspending it in air by only the ropes of the derrick. For the entire duration of the operation, the obelisk will rest on the wedges as it does on the die of the pedestal."

Each worker was wearing an iron helmet to protect him from pieces of iron or wood that could fall from the upper parts of the castle.

While these preparations were executed, the weather became pleasant. The peaceful serenity of the sky seemed to the architect the portent of a happy success.

All of Rome's able-bodied people, men distinguished by birth and talent, had gathered together on St. Peter's Square. The roofs of the houses, and the public buildings that framed the enclosure, were packed with spectators. The enormous crowd pushed into all the streets, all the spots that had a view of the obelisk.

A deathly silence permeated the flowing crowd. All was ready; the men and the horses were only waiting for the signal to start. In that solemn moment Fontana, standing on a platform where he could be seen by everyone, reminded everyone in a strong and sonorous voice of the religious motives that had given rise to the moving of the obelisk.

"The work that we are going to undertake is consecrated to religion, to the exaltation of the Holy Cross. Implore with me the assistance of God, the Supreme Sovereign. Let us ask him for his help, without which all our efforts would be useless." And all those present, nobles, commoners, priests, foreigners, fell to their knees and recited with Fontana a Pater and an Ave.* The architect stood up, waved the flag that he held in his hand, and the sound of the trumpet vibrated in the air for the first time. Everything was set in motion. The capstans turned on their axes, the derricks strained, the ends of the levers lowered, the blows of the sledgehammers resounded dully on the heads of the wedges, and the obelisk, which was leaning two palms toward St. Peter's, returned to perfect verticality. At the sound of the bell, everything became perfectly immobile again. "In this first movement," says the architect, "one would have thought the earth trembled. The castle made a terrible cracking sound."

This noise came from the joints of the structure, and from the ties that had tightened due to the effect of the pressure. But none of the parts had broken. Only the workers who were keeping an eye on the apparatus pointed out the breaking of the first iron band of the cladding at the top. They replaced it by a band of rope, maintained by pins that passed under the base of the monolith. This repair was barely finished when, at a second signal, new efforts were attempted and crowned with complete success. The same signals were repeated twelve times and were followed by the same results. Finally, at the twenty-second hour of the day, the obelisk was lifted 2¾ palms. This height was sufficient for the purposes of the architect. Several rounds were fired from the castle announcing this happy result. All the artillery then issued a salvo to express their happiness. Up till that moment, the workers had remained at their posts; people had brought them food right there.

The morning of the following day was dedicated to checking the apparatus. All the bands of the cladding had broken or twisted. Some had pulled out or cut the abutments that they hit like knives. "This experience showed us," observed Fontana, "that one cannot entirely trust metallic ties. Ropes offer more safety. This is why I took the precaution of tying these fixed points with ropes passing under the bottom of the obelisk."

The cubes of the pedestal were removed and brought to the Holy Father, who repeatedly showed his joy and satisfaction.

After this operation, the obelisk was resting on the wedges that had been inserted under the four lower corners. For these resting points they substituted a platform or cradle on rollers AB. One of its extremities was placed under the bottom, in the space that existed between the wedges. It was on this cradle that the monolith was taken down.

* Nothing is so imposing as to see a large group of people associate the work of human genius with religious sentiment and divine intervention. Here it is not only the learned person and the artist who follow the raising of the monument with interest and love. While the learned and the artists presided with their enlightenment, the people also took part through their vows. Here as elsewhere, Christian worship expresses its eternal and admirable doctrine of equality, and gathers all those present in one prayer. It contains something truly popular and national.

The architect considered this maneuver to be more difficult and much more risky than the first one. It required new settings, a new arrangement within the apparatus.

First of all, it became necessary to insert fixed pulleys in certain places on the face of the cladded obelisk that would be lowered, and next, to move the corresponding machines. Second, four horizontal derricks whose mobile pulleys were attached to the foot of the obelisk were installed. The tops of the tackles were wound around the bells of the four new capstans. Finally, to avoid broken ropes and unforeseen accidents, the architect had someone place, near the middle of the cradled obelisk, a system of mobile props around fixed axes, whose two extremities were surrounded by an iron collar that embraced the three other sides of the monolith. Their purpose was to support the obelisk in the probable case that the operation had to be stopped. To accomplish this, the space between the feet of the props held a winch or mobile roller. Two ropes, attached to the two ends of the cylinders, were each wound several times around its surface and were finally attached to the columns of the castle. With this mechanism, the feet of the props would move away from the axis of the pedestal as the obelisk was lowered toward the horizontal, as long as the winch was free to turn. However, if the rotation was hindered by means of a lever placed in one of the mortises of the roller, then the ropes, now unable to unwind, would maintain the system in a fixed position.

All these preparations were completed on May 7.

As in the first operation, the trumpet controlled the movement of the capstans. The system remained immobile at the first sounding of the castle bell.

The view on Plate XIV and a few words of explanation will suffice to give an exact idea of the process followed by Fontana.

Let us suppose that the forty capstans that surround the body of the obelisk have been properly tightened; the friction, determined by the pressure of the mass on platform AB, would diminish in proportion to the tension transmitted to the ropes of the pulley blocks. According to the settings chosen, this resistance will be easily overcome by the action of the four horizontal pulley blocks. When these machines are at work, the bottom of the monolith will tend to move away from the pedestal. If, at some point, the capstans are released slowly and with caution, the top of the monument will lower toward the horizon while the foot will move away from its original position, taking with it the roller upon which it is resting. Fontana intended to control this last movement in order to always keep the upper part of the apparatus vertical. But it didn't happen this way. The arc described by the top of the obelisk did not correspond to the space covered by the bottom. The result was that the horizontal parts of the pulley blocks, which were restraining the obelisk under a more and more acute angle, became powerful enough to allow the cylinder to slide on the rollers without the help of the horizontal derricks. The speed of this movement had to be slowed down, with the help of a new pulley block that counteracted the first four.

In order to prevent a shock, a jolt, when the load landed on the platform, Fontana had installed at the top of the obelisk five new pulley blocks whose fixed pulleys were attached to the vault of the sacristy of St. Peter's. Those pulley blocks, which Fontana compared to horse bridles, were intended to prevent all unwanted acceleration in the motion of the obelisk.

These various maneuvers—the movement of the arches of the castle, e,e, which had to be removed to permit the free movement of the obelisk—required stopping, which necessitated, up to a certain point, the use of the mobile props that we mentioned before.

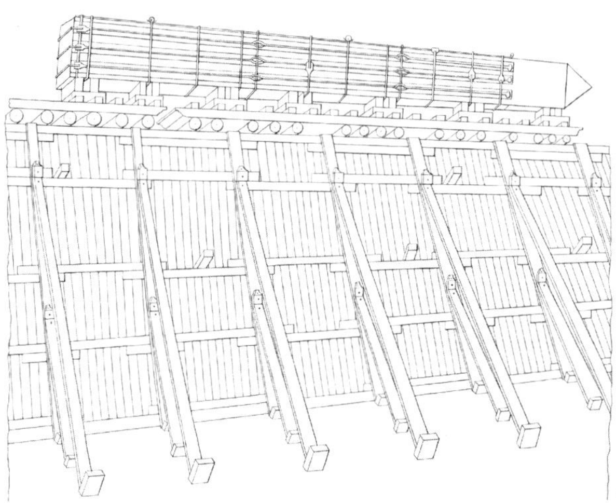

Detail from Plate XIV: The path on which the Vatican obelisk was transported

Finally, at the twenty-second hour, the obelisk was lying on the platform without having experienced the slightest problem.

Sixtus the Fifth was delighted with this first success. "Bravos" were heard from all corners of the square. All of Rome saluted Fontana with its acclamations. The trumpets and the drums of the town accompanied him home, playing fanfares amid the general rejoicing.

The obelisk had just been placed on a roller; now it had to be brought to St. Peter's Square. The straight path that it had to cover was approximately 113 *cannes*.* This transport had to be carried out on a nearly horizontal path. (The difference between the points of departure and arrival was three palms.) This slope, although very minimal, was favorable for the transportation of the load. The path or viaduct, which ended at the top of the new pedestal and surrounded it on all sides, was 37 palms high, over 100 palms wide at the base, and 50 at the top. But around the pedestal, the last two dimensions were 125 palms and 95 palms.

The path was built with sand; an embankment of this kind made with freshly dug soil could not have withstood such a considerable weight without lateral reinforcement. This reinforcement was made of joined planks placed on the mound, three courses of horizontal girders, against which leaned buttresses at certain

* Translators' note: A *canne* was a measure of distance equal to about two meters (six feet).

intervals, each supported by two struts. The intermediate spaces were consolidated by transverse and oblique beams that connected the two sides of the path. In the part corresponding to the pedestal, a framework was set up, of which the cross-section on Plate XIV shows us the details. It was intended to support the bottom of the platform of the castle.

Once this task was agreed upon and carried out, the five supports of the old pedestal were extracted, the apparatus was dismantled, the framework was demolished, and all this material was transported to Vatican Square. Medals, similar to the first ones, were thrown into the new foundations. Two of the medals, struck in gold, had Pius V on one side and the images of Justice and Religion on the other. Between two of the three pieces of marble that composed the socle or the lower support, a plaque of the same stone was inserted, on which were engraved in Latin the name of Pope Sixtus the Fifth, that of Fontana, that of his native town, and a brief summary of the operation, forever memorable, executed by this architect. The medals with the image of Pope Sixtus the Fifth were placed on the top level of the socle.

All those preparations required quite a bit of time. It was then necessary to superimpose the blocks of the pedestal and frame it within the huge castle that had already served to lower the monolith, and to complete the final details of the towing path. It was only after these works were completely finished that the obelisk could be moved. It had remained on its bed of rollers, placed on the same level as the top of the new base. It was advanced toward St. Peter's Square with the help of four apparatuses that included the same number of capstans. It was dragged this way to its final destination, that is to say, with the pyramidion aligned vertically with the axis of the pedestal (Plate XV). Because of this, the platform or bed of rollers that supported it was placed between the last support and the obelisk.

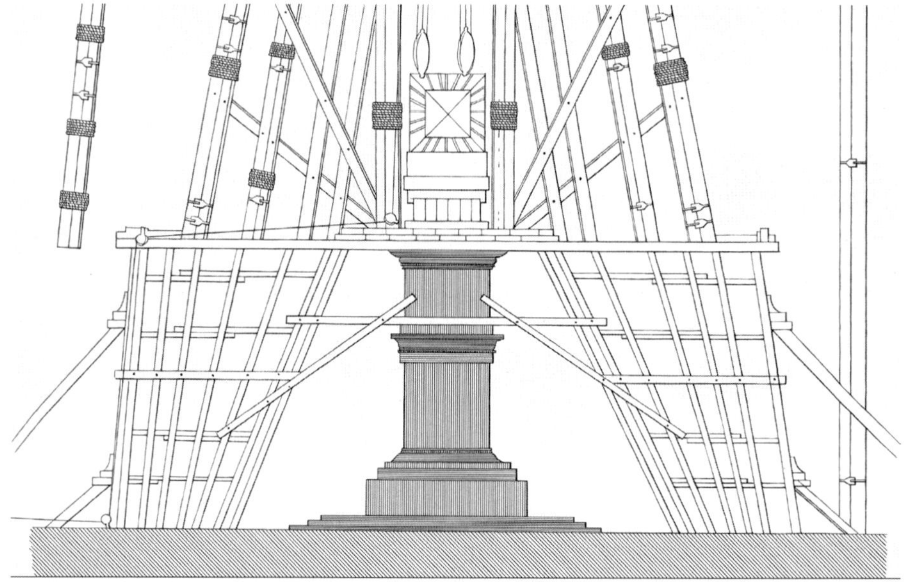

Detail from Plate XIV: Cross-section of the path constructed by Fontana to transport the obelisk to Vatican Square

The pulley block, the machines, the men, and the horses all were set and distributed as in the first operation. Only the four horizontal derricks attached to the base were set up so as to operate in the opposite direction. They were meant to bring the base of the monolith onto its resting bed.

There are but a few things left to say about the erection, which took place on September 10, 1586. This operation was preceded by the same religious ceremonies and preparations. It was carried out by a process similar to the one we described for the lowering of the monolith. Plate XIV will again help to explain it. As the top of the obelisk, by the united effort of 40 capstans moved by 800 men and 140 horses, was rising into space, the horizontal pulley blocks, under the action of four similar machines, caused the base and the platform supporting it to move toward the axis of the castle. The speed of this movement was regulated in such a way as to conserve, as much as possible, a reasonable vertical direction of the ropes of the upper apparatus.

When the axis of the monolith reached a forty-five-degree angle, the workers, exhausted and extremely hungry, asked to stop the maneuver to catch their breaths. During this time period, the monument was resting on the props whose foot was fixed, because of the lever that had been introduced into the winch. Once they had rested, they went back to work. The process continued. After fifty-two movements, the obelisk was rising vertically on its pedestal, but its base leaned on the rolling platform, which it had dragged along during the rotation.

The enthusiasm reached its height when the people saw the obelisk standing on Vatican Square. Some artillery shots heralded this great event. All the people shouted with joy and saluted the architect with unanimous applause.

The operation took from dawn to dusk, about thirteen hours. During all that time, the spectators remained on the square without taking any sustenance.

The following day, they proceeded to remove the roller platform on which the obelisk rested. It was done as we described it (page 187) with forty derricks arranged on the shaft and bearing the same number of capstans, with two systems of levers and wedges inserted under the parts of the base that were not covered. The obelisk, pulled at the same time by the action of all those machines, was resting only on the corners. This allowed them to free the bed that was underneath, and to insert the cast dies at the four angles of the space that it occupied on the top surface of the pedestal. Everything being in place, the obelisk was lowered on these metal supports, by slackening the derricks and levers little by little as the wedges were removed. This difficult work demanded much care. It had to be done slowly with order and deliberation. Therefore, it was not until eight days after the erecting that the obelisk was resting in perfect equilibrium on its four points of support.

Such is the faithful rendering of the facts told by Fontana in the work noted. If there had remained any doubts in the minds of the readers about the assertion that we expressed, these facts would be sufficient to prove the falsity of this historic cry, "Wet the ropes!" The obelisk was never hanging entirely from ropes. It was constantly resting on its base or on one of its lower sides. Otherwise, the tearing of the ropes and other possible events during the lowering and erecting could have caused the obelisk to fall. Fontana had taken care to provide the monolith with a system of props that were intended to maintain it in any position: *Acciò ché essa guglia non posasse mai su le corde, ma restasse sempre appuntellata* ("Although the needle never rested on ropes; it was always supported by props").

Appendix

Overview of the mechanical arts of ancient and modern people — Calculations relative to lowering and erecting the Luxor obelisk

Overview of the Mechanical Arts of Ancient and Modern People

Erecting the Luxor obelisk has given rise to research on the procedures followed by the Egyptians for similar operations. Before completing this work, it may be helpful, with regard to this subject, to glance at the conjectures that gave birth to this spirit of investigation.

It is probable that then, as today, they used machines that allowed the movement of extremely heavy masses with relatively little effort. But we do not know what methods the imagination suggested to these people, and what experiences led to the discovery of this power. Let it be said, we have on this matter neither certainties nor probabilities. If it is true to say that history of the past is in general quite obscure and often imagined, then it is a given that the history of mechanics in particular is completely unknown.

In the absence of written history one might hope to find, in the immense number of objects sculpted or painted on the edifices of this ancient country, the representations of some mechanical apparatus. Again, on this point, our hopes have not been fulfilled. Egypt shows to the traveler the results of her works: the obelisks, columns, colossi, entire rocks transported, raised in major cities from Aswan down to Alexandria. But she has left no information about the machines that were used, no trace, no elements of a science that probably helped them to erect these monuments, objects of their glory and veneration. People have tried to make up for this silence of history by conjecture, by more or less plausible deductions that have created two diametrically opposed systems.

According to some, the Egyptians represented in the temple and tomb bas-reliefs all the emblems or attributes of religion, of science, of the arts, of agriculture and commerce, of navigation, of industries, and of crafts of all kinds. One even sees recreational games of both children and adults. One can see the tools and the attitudes of the workers using them. And nevertheless, one would look in vain in the great number of representations for some clue of what we call mechanical art—any apparatus whatsoever intended to lift or move a heavy burden. What should we conclude from this silence? Should we say it has been overlooked, by negligence? But all of these tableaux are not from the same era. Would something that had been overlooked by the first artist also have been overlooked by those who came after him? These drawings have not been traced one from the other. There are differences between them in the arrangement, in the movement, in the configuration of the sculpted objects on all the tableaux. Is it that they had nothing to represent, nothing to transmit to posterity? And should we not conclude, along with Diodorus Siculus, that these operations were made by brute force with the help of mountains of sand; that it was the work of manual labor, a work without any standard process and varying according to the location? The facts are the proof, say the others: the Egyptian monuments exist, the colossi and the obelisks still rest on their pedestals. These results prove the superiority of the methods that produced them and give a very good idea of the knowledge these people had of the mechanical arts. Moreover, they add, how can one believe that in an industrial country, whose population was never more than seven million people, they did not try to invent machines to save manual labor?

Finally, taking this conjecture still further, some enthusiasts, struck by the grandeur of the ancient monuments, maintain that the Egyptians had discovered all of nature's secrets; that for them, erecting an obelisk was child's play. They tell you how they were able to prove that all our inventions and discoveries, even the most modern, were known to the ancients and written in the books in the famous library of Alexandria. If you were to believe them, you would be tempted to believe that the flotilla of Sesostris was towed to the open sea by steamboats.

These writers base their opinions on material facts whose truth nobody contests. But this truth, considered from the viewpoint of the causes and means that produced it, is nonetheless a conjectural truth, a simple induction, because the causes and means are unknown to us.

Regarding the silence of tradition on this point, should we conclude with the first viewpoint that the Egyptians didn't know the mechanical arts? Assuredly not; and yet, without adopting the very extreme other position, one must recognize the existence of science where one finds its effects intelligently combined. We are forced to look somewhere else for the reason for this silence. Perhaps it can be explained by the political and religious mindset of these people. In Egypt, the priests reserved for themselves the right to keep, read, and interpret the sacred books, in which was found all knowledge revealed by the gods themselves. These men, who seem to pose as demigods on earth, had to enthrall the ignorant population subjected to their power with the help of a few so-called religious beliefs, proven by so-called miracles such as the sounds of the statue of Memnon. And one can presume that the priestly caste, the sole depository of science at that time, the only ones initiated into its mysteries, who took pains to keep this information from the multitudes, had decided to exclude from the symbolic representations traced on the monuments all that could have explained the phenomena that it produced, and thus could have damaged the priestly power by diminishing its prestige. But this, I must admit, is only a hypothesis. To support this assertion, I have no convincing proofs, which, in matters such as this, seem to me indispensable. What strikes me most of all is that experience must have suggested to men constantly working on such enormous loads the means to facilitate their transport and placement. This probability becomes a certainty if one compares the dimensions of these blocks with their weight. It would be physically impossible to use the necessary number of men to move these masses; even if this difficulty did not exist, how could you put a thousand men to work without the help of some apparatus? Thus, since the Egyptians erected gigantic monuments, it means that they employed in these sorts of operations more or less ingenious processes. But we cannot share the opinions of those who favor the idea of ancient machines at the expense of ours. Their reasoning consists in saying that in a country where the population was not very large, they had to bring the highest degree of perfection to the instruments that were designed to minimize the number of people needed for the work. This assertion seems to me quite contestable. In effect, can a nation who, for no useful reason, undertakes to build so many monuments of simple lines, whose number, size, strength, luxury, perfection, decoration, multiplied ad infinitum, surpassing all that the rest of the ancient world left us in this genre and all that the modern world was able to produce—can such a nation be regarded as a people who seek to minimize the number of workers? Isn't it obvious, on the contrary, in view of this prodigious work, that its main program, its essential purpose, was to keep them busy?

Next, the human mind judges, and can only judge, by comparison. To appreciate the superiority of one object over another, it puts them in parallel; it examines their respective properties, the degree of perfection, the advantages and disadvantages that could result from their use and their application. However, in the case that concerns us, the means used by the ancients being totally unknown, how are we to determine which is better? One cannot establish a parallel between what is and what is not. The nothing placed on the pan of a balance scale does not disturb the equilibrium.

To these observations one opposes facts; one cites the colossus of Ozymandias, weighing about eight hundred tons, which was transported from the cataracts of Aswan to Thebes. Here is the most striking example that ancient science offers us.

The objection is not really serious, and the conclusion that one draws seems to me to hardly conform to the rules of a healthy logic. I could cite such and such operation performed on a heavy load with the help of processes that are basic. But to get directly to the point, I will point out that we cannot know whether the apparatus, with the means at its disposal, would stand up to the difficulties of such an enterprise.

Permit me to add to this silence and to demonstrate that people did move, and are still moving, masses more formidable than that of Ozymandias, which, in the system that I refute, would infallibly take us to the conclusion that modern science is superior to that of the ancients.

Without speaking of the rock of St. Petersburg, the raising of the Alexandrian column, the lowering and erecting of the obelisk of Luxor, which proved at least that modern science could undo and redo the work of the Egyptians, we will only cite one example taken from the sea.

It is not a colossus of eight hundred tons, but a vessel of 2,500 tons, that naval art will seize in the middle of the sea and haul up a plane inclined at twelve degrees to the horizon, in a few hours. Sixteen capstans that act directly on as many cables attached to the vessel all accomplish this immense work. I doubt that one could imagine a machine that is simultaneously simpler, more powerful, and more economical.*

Now, if you study the construction of this floating fortress, the operation of launching, the yet more remarkable operation of careening, which consists in inclining this enormous mass to gradually immerse all the parts of the hull all the way to the keel; if you think about how one raises the topmast and the topgallant mast, how one maneuvers the anchors, the small boats, and all the weight of the armory; if you follow this same ship to the middle of the vast ocean where it is victorious against the waves and storms with the help of its immense sails, which are spread to the winds or removed from them with a speed that would astonish a stagehand at the opera; if you consider, moreover, that all these maneuvers are executed on an object tormented by the elements, you will be brought to believe that a vessel must be full of very complicated machines to conquer so many unforeseen obstacles, and to produce effects that defy the imagination. However, it contains

* It is to Mr. L'Eveque, marine engineer, that we owe this apparatus that was used to haul the three-bridged vessel called *Le Majestueux*.

only derricks, capstans, masts, and ropes. These instruments are sufficient for all the operations that must be carried out in a multitude of variable circumstances.

The simplicity of the means, the greatness of the results, the speed of execution, all combine to prove that naval art, the science of measures and calculations, has reached a very high degree of perfection. In the presence of such facts, and a thousand others that I could cite, how are we to decide the preeminence of ancient or modern? Isn't it more rational to conclude that the mechanical arts are not a science of new creation, that they have been cultivated and practiced in the most remote centuries, and that, like all that God has made for the use and development of the human race, the mechanical arts passed through a series of successive advances? But it is impossible, and truly regrettable, to present a comparative table of the mechanics used in different times and places to move tremendous masses. We lack historical sources. We find traces of them, however, in Vitruvius and in Ammianus Marcellinus. The first tells us that pulleys, capstans, and cranes were known in his time. The second gives us an account of erecting an obelisk with the help of the same equipment. So, at any known time, humans have used levers, derricks, and capstans in large operations. These instruments are simple, powerful, economical, easy to assemble and fix, and to apply specifically on a particular part of the mass, and in a given direction. Such are the advantages that must have convinced the practitioners to use them in large operations, where exceptionally heavy masses have to be moved with a relatively small effort.

When one has realized, through calculation or experience, the power of these machines used alone or combined, it seems at first that nothing is easier than to move the heaviest loads. It is, indeed, very simple on paper, but it is not the same in practice. The moving of heavy masses is always a difficult enterprise. The theoretician reasons on the basis of hypotheses; he supposes that all parts of the apparatus are indestructible, and indestructibly tied to fixed points, that the planes on which he operates are not deformable, that movement is uniform, etc. However, this is pure fiction. When it comes time, if I may so put it, to put the hatchet to the wood, to put into practice all these hypotheses in the field, at least as much as possible, a thousand unforeseen obstacles appear that force one to modify, or even abandon, a project whose success did not seem in doubt.

It is only when you are on stage, in the field and in the presence of the objects, that you can appreciate these difficulties. They do not precisely exist in the conception of the idea, which in certain cases is everything, but rather in the details of the execution. The question is first to create fixed points able to resist the weight of the mass, to adapt the machines to the locations, to the mode of action that needs to be produced, and to take the necessary measures to avoid wasted effort. Then one needs to determine the absolute resistance of the various parts of the machines that definitely would be shattered were they not composed of the best materials, made with precision and of ample dimensions; to take into consideration the compression of the ground, the compression of the wood, the elasticity of the system, and a multitude of variable circumstances that can only be calculated with the help of experience. One realizes how difficult these details are and how much influence they have on the end results. It is because of a lack of studying this practical part of the science that some theoreticians have sometimes failed in their enterprises.

"Theory without practice," said an ancient author, "is like a vessel with neither sail nor helm; it remains motionless; one could not sail the seas without being shipwrecked. One must also realize that practice without theory is also like a vessel without a compass, without nautical instruments, which could not navigate without getting lost."

It is agreed that theory and practice are equally necessary to construct a good machine. The theoretician is stopped at every step because of lack of data regarding the quality, the resistance of the materials, and the details of construction and installation. These data result from a series of observations and experiences, which talent and insight could not replace. On the other hand, there are general principles that are indispensable to the practitioner, and without which, the use he makes of the machines is an uncertain action, an attempt that success alone can justify, whereas calculations give him the certainty that his goals will be achieved. Imagination and practice are excellent guides to invent a machine, but the machine, once put together, must be submitted to rigorous testing of the theory in order to guarantee the results. In a word, theory must be based on practice. The latter supports it and serves as a base; the former perfects it and gives it certainty.

To leave nothing to chance is to lack courage, according to some; according to us, it is prudence, especially when you have to raise an obelisk brought at great expense into the heart of a capital, to perform in the presence of thousands of spectators a maneuver whose failure would bring about terrible accidents, the loss of several hundred workers.

This is how we had to proceed in the three operations: to take down, to transport, and to erect the obelisk of Luxor.

I will briefly explain the principles on which rests the applied theory of the machines we used; I will then give a quick overview of the calculations which showed, in mathematical terms, the strength of the forces. With the help of these results, I will lead the reader to conclude with me that we mastered the operation, that the greater or lesser tractions or pressures, supported by ropes or other parts of the machine, were insufficient to break the machines or pull out the fixed points to which they were attached.

Calculations Relative to Lowering and Raising the Obelisk of Luxor

The dry details that we will present will inevitably force us to use technical words, which must be clearly defined to understand the text.

Let *p* and *p′* be two pulleys with pulley wheels (Plate XV, figure 3). The rope *cdba* which surrounds the wooden box in which the pulley wheels turn is called a grommet. The band *ab*, which is part of it, serves to attach the pulley to a fixed point or to the resistance.

This said, if the end of a long rope is attached to the heel, *b*, of the lower pulley, and the other end is threaded alternately on each of the two wheels of the two pulleys, going successively bottom to top, and from right to left, the result will be a machine that the person on land calls a *derrick* and the sailor a *pulley block*.

In the navy, one calls the ropes between the pulleys *tackle*. The one attached at *b* has the name *fixed*. The part of the rope that comes out of the last pulley wheel is improperly called the *runner* or *puller*. The runner doesn't *run* any more or less than the other tackles, but it is upon it that the motive force acts, with the result, as we will show, 1) that the movement and power of a pulley block depend on its action; 2) that the cord holds the heaviest load and controls that of the others. It is thus with reason that Fontana called it *chief* (*il capo*) of the tackle. We will adopt this appellation, which seems to us more rational.

The pulley blocks are distinguished by the number of ropes they hold. They are called a four-, five-, or six-branch pulley block. The strongest and biggest are called heavy pulley blocks.

The function of a pulley block is very simple. Let us suppose that we need to move a mass M from A to B with the help of this apparatus. After being attached to the pulley *p′* at a point C situated in the direction AB, one attaches the resistance to the grommet on the other pulley. Then, all that is needed is to exert on the

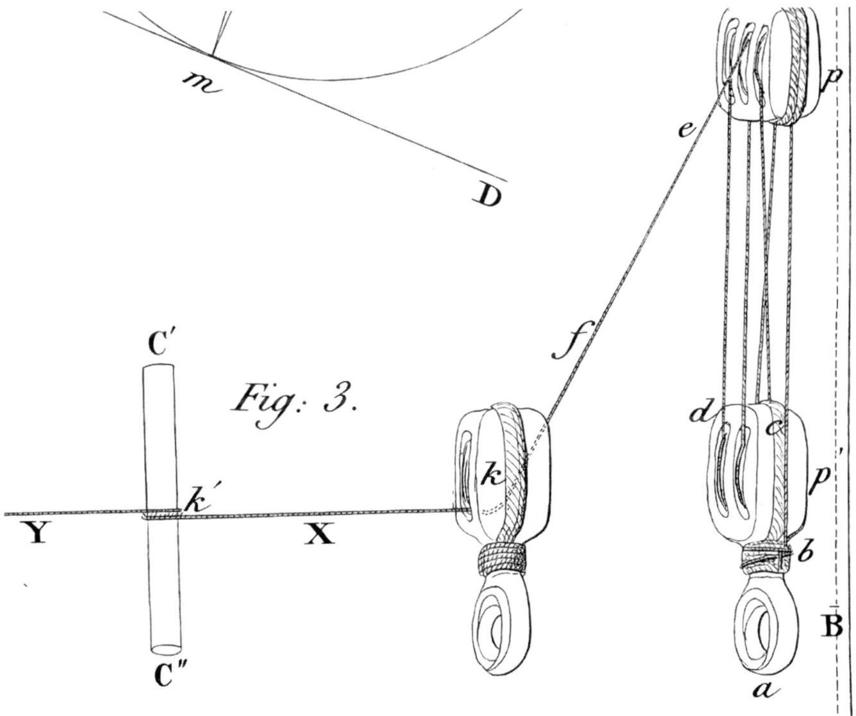

Detail from Plate XV, figure 3: Pulleys with pulley wheels

chief pulley *ef* a tension sufficient to unwind the tackle, which determines the progressive movement of the mass toward the place where we want to bring it.

There is in mechanics a principle known as *the principle of virtual speeds*, which can be translated in ordinary language as follows:

Whatever the apparatus put into play—levers, capstans, derricks—whether we use them on their own or in tandem, there is always compensation between the intensity of forces and the distances covered in the same time by their point of application. Because of that, the small force has to work hard, to give very little effort to the heavy mass against which it is fighting. Thus one effort of 10 kg can triumph over a weight of 10,000 kg, but only on the condition that it moves and covers as much as 1,000 times the distance that the weight moves. Here is the whole secret of the effects produced by these mechanical machines: they allow extremely heavy masses to be moved with a relatively weak effort, not because they multiply the power of the men working them, but because they transpose it into a new form. Theoretically thinking, the result is always the same; it is either force or space, or time and force, which balance each other reciprocally. Indeed, when you build a six-branch pulley, for example, the result is that a man who exerts a force of 10 kg on the chief pulley can overcome a resistance six times greater attached to the mobile pulley *p*, because the distance covered by the mass will only be 1/6 of the length of the rope unwound by the motive force. In general, the force transmitted to each of the two pulleys is equal to the force that acts on the chief pulley, repeated as many times as there are ropes on the pulley that has been put into motion.

Let us suppose that the chief of this same pulley is attached to a capstan moved by 50 men, each exerting a force of 10 kg at the ends of the bars, 500 kg in all. The circle that they will describe will be 10 meters in diameter, for example, whereas the chief pulley will describe a circle of 2 meters in diameter if such is the diameter of the body of the capstan. The mass attached to the pulley, *p*, will only move, as we have said, 1/6 the length of the rope wound around the capstan, which corresponds to 1/30 of the distance covered by each of the men. This means that 50 men, whose combined efforts equal only 500 kg, can overcome a resistance 30 times greater, that is, 15,000 kg, with the help of a capstan and a derrick.

These theoretical results are far from being confirmed by experience, the reason being that the various parts of our apparatus, whatever it is, cannot be put into play without more or less considerable loss of force.

In a derrick, part of the force is absorbed by the stiffness of the ropes and by the friction of the wheels on their axles. These resistances keep the chief pulley from fully transmitting its action to the second rope. It is the same for the latter (second rope) and with respect to the next, and so on and so on. So the tensions of the tackle decrease progressively from the chief pulley up to the sleeper (fixed pulley), which supports the least traction.

The result of numerous experiments made by Coulomb shows that the resistance that a rope offers as it winds around a cylinder is composed of two terms. The first expresses the effort needed to bend the rope and wind it before it is taut. This effort, being very small compared to the force on the rope, can be ignored without any consequences. The second is proportional to the product of a certain force of the diameter of the rope, by the tension it supports, and inverse to the diameter of the cylinder or of the wheel on which it is wound. This theorem translated algebraically gives the following formula:

$$\frac{\alpha d^{\mu}}{D}\cdot T \left\{ \begin{array}{l} d \text{ diameter of the rope} \\ D \text{ diameter of the cylinder or wheel} \\ T \text{ tension of the rope} \\ \alpha \text{ and } \mu \text{ constants whose values depend} \\ \quad \text{on the thickness and nature of the rope} \end{array} \right.$$

Starting from these data, it can be demonstrated that, in a pulley block in equilibrium, the tensions of the tackle form a decreasing geometric progression, from which one deduces the two equations:

$$(1)\ T_n = \frac{q^{n-1}(q-1)}{q^n-1}\cdot M \qquad (2)\ T_1 = \frac{q-1}{q^n-1}\cdot M$$

T_n is the tension on the chief pulley, T_1 that on the sleeper, n the number of ropes attached to the mobile pulley, M the mass to be moved, and q the ratio of the progression representing the quantity:

$$(3)\ \frac{D+d+2r f'+\alpha d^{\mu}}{D+d-2r f'}$$

in which $\frac{f}{\sqrt{1+f^2}}$ has been replaced by f', where f represents the ratio of the friction to the pressure that the wheels exert on the axle, D the diameter of the wheels, $2r$ the diameter of the axle, d the diameter of the rope, and α and μ the constants of which we just spoke.

If the chief block is attached to a capstan moved by men whose combined efforts are represented by a force F, we will have for the equilibrium equation:

$$(4)\ 2RF = (D_1+d+2rf'+\alpha d^{\mu})T_n$$

R is the radius of the circumference described by force F; D_1 the diameter of the capstan at the level of the unrolling; the other letters are as in the preceding formulas.

Let us apply these equations to the machines that were used to move the obelisk.

Force of the Motors

Coulomb found that for a clean, new rope of 0.02 m diameter and larger, one can make $\mu = 2$ and $\alpha = 24$ in the formula $\frac{\alpha d^{\mu}}{D}\cdot T$ which expresses the resistance offered by a rope at its point of winding on cylinder D. As for the value of f, or the relationship between friction and pressure exerted by an iron tree in copper boots, experience showed that it was equal to 0.133.

Pulley block	Diameter of all wheels	$D = 0.05$ m
	Diameter of the drum around which they turn	$2r = 0.05$
	Relation between friction and pressure	$f = 0.133$
	Number of ropes on the mobile pulley	$n = \begin{cases} 6 \\ 7 \end{cases}$
	Diameter of those ropes or tackles	$d = 0.05$
	The tackle being clean and new, we will have:	$\begin{cases} \mu = 2 \\ \alpha = 24 \end{cases}$
	Substituting these numbers in the equations (1), (2), and (3), one finds q = 1.186 and	$\begin{cases} T_6 = 0.245 \cdot M \\ T_1 = 0.104 \cdot M \end{cases}$ $\begin{cases} T_7 = 0.225 \cdot M \\ T_1 = 0.080 \cdot M \end{cases}$
Capstan	Diameter of capstans at the level of winding	$D_1 = 0.80$
	Average diameter of iron axle of this machine	$2r = 0.13$
	For everything dealing with the friction and tension of the ropes as above	" "
	Substituting in (3) gives	$2RF = 0.927 \cdot T_n$

These calculations produce the following results: First, that the force to be exerted on the chief pulley of a pulley block with 6 ropes to overcome a resistance attached to the mobile pulley is equal to 0.245 of this resistance, whereas if the friction and tension of the ropes are not taken into account, it would only be 1/6 of this value. Second, that the tension of the chief is more than double that of the sleeper.

This inequality of tension on the tackles produces different pressures on the axis of the pulley; the part of this tree which corresponds to the wheel upon which the chief runs supports the strongest load. It is for this reason that in practice the pulley system is set up in such a way that this rope is always attached to the wheel in the middle of the top pulley.

Let us now suppose that, pulley p′ being fixed, a force M acts upon the upper grommet in order to move pulley p′ away from the resting point by making the tackles unwind from bottom to top. As in the preceding case, this force will spread unequally on these ropes, whose tensions will again form a geometric progression; but the progression will decrease, starting from the sleeper, in such a way that the chief will support the *minimum* traction. We will have for the equilibrium equation of a pulley system of 6 ropes:

$$(4)\ T_n = \frac{q-1}{q^n-1} \cdot M = 0.104 \cdot M$$

$$(5)\ T_1 = \frac{q^{n-1}(q-1)}{q^n-1}\ M = 0.245 \cdot M \text{ (see table below)}$$

If the chief passes by a return pulley *k* which changes its direction at a right angle, the friction of the wheel on the drum and the tension of the rope will prevent its action from being entirely transmitted to the rope kk′.

Given tension X in the rope kk′, we will have:

$$T_n(D+d) = X(D+d+\alpha d^{\mu}) + 2rf'\sqrt{T_n^{\,2} + X^2}$$

Substituting for D, d, α, μ, r, f', their numerical values taken from the table above, we have, after all reductions:

$$(6)\ 0.85\ T_n = X$$

If this same rope is rolled in a double turn on a fixed cylinder C′C′, and then turns again and passes pulley R′ to another fixed cylinder VV, on which it turns in a single turn, this combination of friction and resistance due to the tension of the ropes will further reduce the action transmitted by the force of the extremity u of this rope.

We designate by Y the tension of the part of this rope comprised between C′ and R′, by Z the tension of the portion that goes from the return pulley to the fixed cylinder VV, and by u the force exerted on its extremity. We will have successively:

$$0.85 \cdot T_n = X^{*}$$

$$X = Y\left(e^{\frac{ms}{r}} + \frac{\alpha d^{\mu}}{2r+d}\right)$$

$$Y = \frac{Z}{0.85}$$

$$Z = u\left(e^{\frac{ms'}{r'}} + \frac{\alpha d^{\mu}}{2r'+d}\right)$$

---

* Let T_1 and T_0 be the tensions of the extremities of a rope which coils around the fixed cylinder 0, T its tension in any point A, fig. 4.

At the point infinitely adjacent to point A, T will become T+dT because of the friction resulting from the pressure of the rope on the infinitely small arc ds=AB. Through each of the points A and B let us draw a tangent to the circle, and let us join CO, in both right angle triangles CAI and IAO, we will have:

CI: AC :: AI: OA
2C: AC :: 2AI: OA

Thus 2CI :T ; ds : r=OA and thus 2CI where the pressure on the arc ds is equal to $\frac{ds}{r} \cdot T$, m being the ratio of the pressure to the friction; the resistance due to the pressure which we represented by dT will give the equation $dT\ \frac{mds}{r} \cdot T$

Integrating we find:

$$\int \frac{dT}{T} = \int \frac{mds}{r} \quad \text{Or} \quad \log. T = \frac{ms}{r} + C$$

For s=0 $T=T_0$ thus log. T_0=C, and thus $T_1 = g^{\frac{ms}{r}}\ T_0$

Multiplying these equations member by member, we obtain the final result:

$$(7)\ (0.85)^2 T_n = \left(e^{\frac{ms}{r}} + \frac{\alpha d^{\mu}}{2r+d} \right) \left(e^{\frac{ms'}{r'}} + \frac{\alpha d^{\mu}}{2r'+d} \right) u$$

where m is the ratio of the friction to the pressure that the rope exerts on each of the fixed cylinders; r and r' their average radius; s and s' the wound arcs; e the base of the Napierian logarithms; the other letters as above.

The quantity m takes different values according to the form and nature of the wood on which the rope is wound. If the wood is tender and soft, like what was used in Luxor, the rope, subjected to a strong pressure, penetrates inside, digging into the surface of the uneven grooves, which increases the friction, especially if the piece is not perfectly round. These considerations led us to determine through direct experience the value of this coefficient. We found that in the most unfavorable case it did not surpass 0.222. The results deduced from this number will thus be too strong rather than too weak. Substituting in (7) and observing that $s = 4\pi r$, $s' = 4\pi r'$, $e = 2.718$, and $n = 6$, we have for a 6-branch pulley:

$$(8)\ u = 0.001084 \cdot M$$

Thus, for a pulley of 6 branches, set as we just indicated, if one pulls from top to bottom, if the strap of the pulley p is pulling a weight of 20,000 kg, for example, this force will only transmit to the extremity of the chief pulley a force of 21.68 kg, that is to say, about 1,000 times smaller; the rest will be absorbed by the friction and the tensions of the ropes.

With the help of these formulae, it is easy to answer all the questions relative to moving the loads. These are of two kinds. If it is a question of displacing a mass without lifting it, by sliding it along a horizontal plane, the only resistance to overcome consists of friction and adhesion of the surfaces in contact. These resistances can be considerably reduced by greasing the bodies or by carrying the load on rollers, as in the moving of the Vatican obelisk, or by using balls placed in grooves, as with the rock of St. Petersburg. With the help of the same artifice, that is to say, by substituting, for the friction of the wheels on stone or earth, that of a band of iron, we succeeded in reducing the friction of a load that is 1/16 of its weight on its natural path to 1/200.

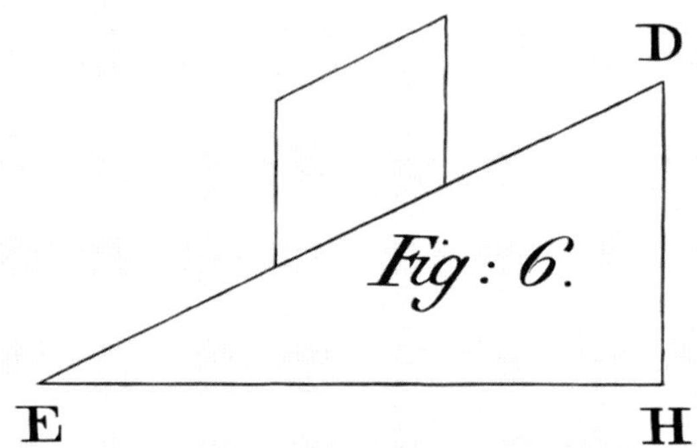

Detail from Plate XV, figure 6: Parallelogram illustrating the forces exerted on the ropes and lowering apparatus as the obelisk was being lowered

If the surface upon which the body glides is inclined, one will have to overcome the resistance due to the friction and adherence, plus a small portion of the weight, which will be proportional to the slope. Thus let M be a mass placed on an inclined plane DE, Plate XV figure 6, DH = h, and HE = b. We will have for the parallelogram of the forces:

$$R = \frac{h \cdot M}{\sqrt{b^2 + h^2}}$$

R is the product of the mass parallel to the inclined plane DE.

Thus, $\frac{h \cdot M}{\sqrt{b^2 + h^2}}$ expresses the force of the mass as it descends, which, it is easy to see, varies with the slope of the plane DE; if the body is in equilibrium, the equation measures the friction and adherence of the surfaces in contact. Now, we know that a vessel placed on a cradle whose slope is equal to 1/12 will slide on its own, even after resting several days on its cradle, from which we deduced that the vector of the weight following the slope, which is in effect equal to M/12, is sufficient to overcome the friction and adherence. Thus, by taking M/12 as the measure of these resistances, the value that one can deduce will be the upper limit, since the inclination of 1/12 is greater, not less, than that which results in equilibrium. According to these data, the force F to be exerted parallel to the plane DE in order to raise mass M will be expressed by:

$$(9)\quad F = \frac{M}{12} + \frac{h}{\sqrt{h^2 + b^2}} \cdot M$$

If the slope is gentle, the quantity $\frac{h}{\sqrt{h^2 + b^2}}$ will be very small, and as a result, a very heavy load, M, can be moved with a relatively weak force, F. The result of this is that one of the simplest means to lift great masses consists in moving them up inclined planes. Thus, it was by an inclined plane that the obelisk was loaded aboard the ship. It is also with the help of inclined planes that it was transported from the level of the River Seine to that of its pedestal. It was placed on a cradle sliding on a wooden slipway, planed and tallowed like that of a ship. Thus formula (9) can be applied to the hauling of the obelisk. The greatest slope that this monolith had to climb was the ramp of the Pont de la Concorde, which was equal to a 10% grade.

The mass to be moved, including the weight of the sled and the apparatus, was estimated to be 283,636 kg. Using this number in the preceding formula, M = 283,636 kg and h/b = 1/10, the resistance to overcome is equal to 52,000 kg. Thus, to make the obelisk go up the inclined plane was the same as lifting vertically a weight of 52,000 kg, suspended by a rope. Five pulleys with 6 lines tied to the same number of capstans, each moved by 48 artillerymen, were sufficient. Indeed, supposing that the force exerted on an average spoke of 2.25 m by each artilleryman is equal to 12 kg, the force of a capstan, calculated according to the formula 2FR = 0.927 T_6

and T_6 = 0.245 M, gives an actual effect of 11,412 kg. The force transmitted to the obelisk will thus be 11,412 kg per capstan, and the 5 capstans together will yield a pull of 57,060 kg. This result can be regarded as a lower limit because a robust man is able to supply a force of 20 kg for several hours, instead of the 12 that I had assumed.

Determination of Weight and Center of Gravity of the Mass to Be Moved

Basic Dimensions of the Obelisk (Plate XV, figure 5)

Side at the base	AB = 2.425 meters
Side at the highest square	ab = 1.54 meters
Height of the shaft	h = 20.9 meters
Height of the pyramidion	h' = 1.94 meters

If at the four sides, *ab, bc, cd, de* of the upper square *abcd*, we imagine vertical planes, those planes will divide the shaft of the monolith into 9 solids, the height of all of which will be that of the shaft, that is the parallelopiped *abcd*, 4 pyramids equal to A*nao*, and four triangular prisms equivalent to *nida*. With V being the volume of the shaft, one will have:

$$V = (abcd \times h) + (4 \times Anao \times h/3) + (4 \times nida \times h/2)$$

Doing the calculations and adding the volume of the pyramidion, which is equal to *abcd* ($h'/3$), the total volume is equal to 85 cubic meters, which gives a total weight of 229,500 kg. (The specific gravity of granite, deduced from several experiments, = 2.7.) With the cladding, the total weight of the mass to be moved is equal to 250,000 kg.

As to the height of its center of gravity, which we designate by X, it can be deduced from the following equations which express the sum of the moments, taken with respect to the plane of the base of the 10 solids of which the obelisk is composed, is equal to that of the total volume:

$$X = \frac{\left(abcd \cdot h \cdot \frac{h}{2}\right) + \left(4 \cdot Anao \cdot \frac{h}{3} \cdot \frac{h}{4}\right) + \left(4 \cdot nida \cdot \frac{h}{2} \cdot \frac{h}{3}\right) + \left(abcd \cdot \frac{h'}{3}\left(\frac{h'}{4} + h\right)\right)}{V + abcd \cdot \frac{h'}{3}} = 9.16$$

Calculations Relating to the Lowering Apparatus, the Holding and Erecting Mechanisms

The lowering apparatus and retaining apparatus being installed as indicated (Plate II), let us imagine a plane Q going through the axis of the obelisk and that of the cladding (Plate XV). This plane, dividing the system into two symmetrical parts, will contain the resultants of all forces. The resultant of the three pulley blocks that tend to the lowering cable is found also in a plane defined by the axis of this rope perpendicular to the first. Their intersection A′B′ will thus determine the straight line. Therefore, the three capstans, with the help of the derricks, pull the top of the monolith toward them as would a singular directional force along A′B′ and applied to A′. At the same time, if by the axis of the upper tie-beam of the cladding (Plate II) we draw three new planes HH′, RR′, and CC′ perpendicular to plane Q, those planes will divide the system of stays, that of the retaining apparatus and that of the cladding, into two symmetrical parts and will cut the first one along three straight lines HH′, RR′, and CC′ (Plate XV). The straight line HH′ will coincide with the direction of the resultant of the stays, the second with the direction of the retaining derricks, and the center of gravity of the cladding will be found on the third one (straight line).

Let us connect the center of rotation *o* to the center of gravity G of the obelisk and to the point A′. We can regard each of the straight lines as two rigid and non-extendable rods build into the pivot of the base, or rather like two bars of a horizontal winch, which would have the wooden cylinder *oc* as its upright. The first one (straight line) is supported at G by the weight M of the heavy object following the vertical GV; the second is supported at A′ by the resultant of the pulley apparatus and by the tension of the stays. The same is true for the straight line R*o*′ with respect to the cylinder *o*′. The straight line is pulled from top to bottom by the weight of the cladding, of which the center of gravity is located at *o*″, and in the opposite direction by the resultant stays. We do not have to worry about the force RR′, which will not be put into action until the moment when the center of gravity exceeds the hinge [pivot point]. Thus, the problem comes back to establishing the conditions of equilibrium of the two winches *o* and *o*′.

Let P be the weight of the cladding, *p* its distance to the center *o*′ of the moment, H the tension of the cables, *h* its distance to the point o′. Let A be the tension of the rope A′B′, M the weight of the obelisk, and *a m* and *h*′ the length of the three perpendiculars lowered from point *o* in the direction of these two forces and of the line HH′. The equilibrium equation of the winch *o*′*d* will be:

$$Hh = Pp + rf'R \qquad R=\sqrt{P^2+H^2+2PH\cos\alpha}$$

R being the resultant of forces P and H; α the angle they form between them

and for the equilibrium of the winch *oc*:

$$Aa = Mm + Hh' + r'f'R'$$
$$R''=\sqrt{H^2+M^2+2MH\cos\beta}$$
$$R'=\sqrt{R''^2+A^2+2AR''\cos\gamma}$$

R″ is the resultant of the two forces M and H which form between them an angle β. R′ is the resultant of R″ and A, which intersect under the angle *f*. *r and* r′ are the radii of the wood cylinders which serve as axes of rotation; f' for $\frac{f}{\sqrt{1+f^2}}$, in which *f* represents the ratio of the friction to the pressure exerted by the rollers in their encasement, these rollers being of oak and greased with tallow; we have seen before that $f = 1/12$.

The weight of each shear that makes up the bridge, including the pulley with three rollers tied to its top, is equal to 2,800 kg, and its center of gravity is located at 7.90 m from the large end. Thus P = 22,400 kg and *o*′*o*″ = 7.90 m.

Replacing in these formulae the other letters by their numerical values on Plate XV, the values for H, R″, and R′, and finally A can be deduced. All calculations done, we will find that A = 23,612 kg. It is easy to see by inspection of the figure to which we are referring that the value will diminish as the obelisk is lowered toward the horizontal. Thus, the *maximum* tension to be transmitted to the lowering cable will be 23,612 kg.

We have found for the force of one of our capstans 2RF = 0.927 (see the force of the machines). Each capstan is supplied with 16 bars, each turned by four men, 64 in all. The average spoke used by each man is 2 meters, thus R = 2. Estimating the effort of one man at 10 kg, the total force is F = 640 kg, from which it can be calculated that the tension of the rope winding around the machine is $T_n = 2{,}761$ kg. Since this tackle passes through a return pulley *m* (Plate II), the friction of the roller on the axle and the tension of the rope will prevent its force from acting entirely on the next rope T_6. The angle that they form between them being very small, it can be ignored; we will have thus (see pages 200 and 201):

$$T_n = qT_6, \text{ thus } T_6 = 2{,}329 \text{ kg}$$

Since the pulley on which this tackle acts carries 6 ropes, let us assume n = 6 in formula (1). We get $T_6 = \frac{q^5(q-1)}{q^6-1} \cdot M$ M=0.245 M=2,329 kg and thus M = 9,506 kg.

Thus, each capstan will pull the mobile pulley of the corresponding derrick as would a weight of 9,506 kg. The force applied to the lowering cable by the three capstans will thus be three times greater and equal to 28,518 kg, a force that exceeds the maximum resistance by about 5,000 kg. If unforeseen circumstances had required a greater force, we would have found it in the same machine, because the force of a man on a capstan bar is actually 15 or even 20 kg instead of the 10 that I had assumed. He can sustain this effort for more than 2 hours.

When the vertical of the center of gravity of the mass has reached the pivot point, the obelisk falling on its own, the role of the capstans will be finished.

Let us consider the system in equilibrium, in one of the variety of positions that the center of gravity will occupy between points G and (5). The last one corresponds to the largest value in the fall of the obelisk.

It is evident that during the rotation the two straight lines *o*G and *o*A′, assumed to be inflexible, will describe in space circumferences of circles that will have as a common center point *o*. The angle that they form between them being invariable, it follows that when the center of gravity coincides with (4), the straight line *o*A″ will be determined by the condition that the angle A′oG = A″o (4). With the distance from A″ to the top of the shear legs being measured by the length of the cables, if an arc of a circle is traced from this point as center with a radius equal to RA′, its intersection (4′) with a circumference described by the point R will determine the position of this last point corresponding to A″. Taking $o'(4')$, $(4')A''$, and $(4')R'$, the first straight line will represent the axis of the bridge, the second one the direction of the resultant of the cables, and the third that of the retaining pulleys, which abut the center of the platform where the fixed pulleys of the derrick are attached, as mentioned above; the problem is to establish the condition of equilibrium of the same two winches. We will have for the first *oc*:

$$Mm = Hh' + rf'\,R_1 \qquad R_1 = \sqrt{H^2+M^2+2HM\cos\alpha}$$

M = weight of the mass H = resultant of the cables
R_1 = resultant of forces M and H
α = the angles they form
h', f', r, m as described

and for the second $o'd$:

$$Hh = Rr'' + r'f'\,R' \qquad R' = \sqrt{H^2+R^2+2HR\cos\beta}$$

R = resultant of the derricks
R'' = the distance to point o'
R' = the resultant of forces R and H
β = the angle formed by them
h and r' as described

Substituting in these formulae, one will deduce the numerical values of H, R_1, R, and R′.

Knowing H or the resultants of the cables, it will be easy to deduce from it the tension of each. There are 16 of them and they intersect, two by two, on the resultant HH′ following angles that one can consider as being equal to their average φ. Thus, designating the tension on each of them by T, we will have:

$$T = \frac{H}{8\sqrt{2(1+\cos\varphi)}} = \frac{H}{8.2\cos\frac{\varphi}{2}}$$

The restraining pulleys being parallel among them, the mobile pulley of each derrick will be pulled from the bottom to the top by a force expressed by R/8; thus, according to formula (8), the force transmitted to the extremity of the head pulley will be equal to 0.001084(R/8) (see pages 202 and 203).

By doing the same thing on the raising apparatus as with the previous one, we obtain figure 2, in which AB represents the direction of the resultant of the derricks which we will designate by R, BC that of the resultant of the cables, Bo' the axis of the bridge, and $o(1)$ the line that joins the center of gravity of the mass to the center of the movement. When the center of gravity coincides with (4), each of the lines just mentioned will have a position determined by a construction similar to the one we did on figure 1, and the system will be determined as in the previous case by establishing the conditions of equilibrium of the winches *o* and o'. The first system is brought out in (4) by the weight of the mass acting vertically at the extremity of the lever *o* (4) and in (4′) by the tension of the cables. We will thus have for the equilibrium of these forces:

$$(a)\quad \begin{cases} Hh' = Mm'' + rf'\,R_1 \\ R_1 = \sqrt{H^2+M^2+2MH\cos\beta} \end{cases}$$

M = weight of the mass
H = resultant of the cables
R_1 = resultant of forces M & H
β = the angles that they form
r = radius of winch *om*
h' and m are the length of the perpendicular lowered from point *o* on each of the forces M and H

In the same way, the equilibrium of the winch *o′* will give us the equations:*

$$(b)\quad \begin{cases} Rr'' = Hh + r'f'\,R' \\ R' = \sqrt{R^2+H^2+2RH\cos\alpha} \end{cases}$$

R = resultant of the derricks
R′ = resultant of the forces R & H
α = the angle formed between them
r = radius of cylinder *o′*
r'' and h = length of the perpendicular lowered from point o′ on H and R
f' for $\frac{f}{\sqrt{1+f^2}}$ in which *f* represents the ratio of the friction to the pressure

The equations (*a*) will produce the value of H and R, that is to say, the resultant of the cables and the pressure exerted on the pivot of the obelisk. Substituting in (*b*) we will deduce R and R′. The first force represents the force to be transmitted to the apparatus to overcome the resistance, the second the pressure supported by the base of the cladding. Once these quantities are known, it is easy to deduce the tensions of each cable or tackle, etc., etc.

Such are the calculations that we had to make to appreciate at their true value the intensity of the forces that had to be put into play, as well as the forces of traction or the tension exerted on the fixed points and ropes of the apparatus. These calculations are long and

* We did not take into consideration in these calculations the weight of the cladding, which can be ignored without consequences; as a matter of fact, its force is favorable to what we are trying to do.

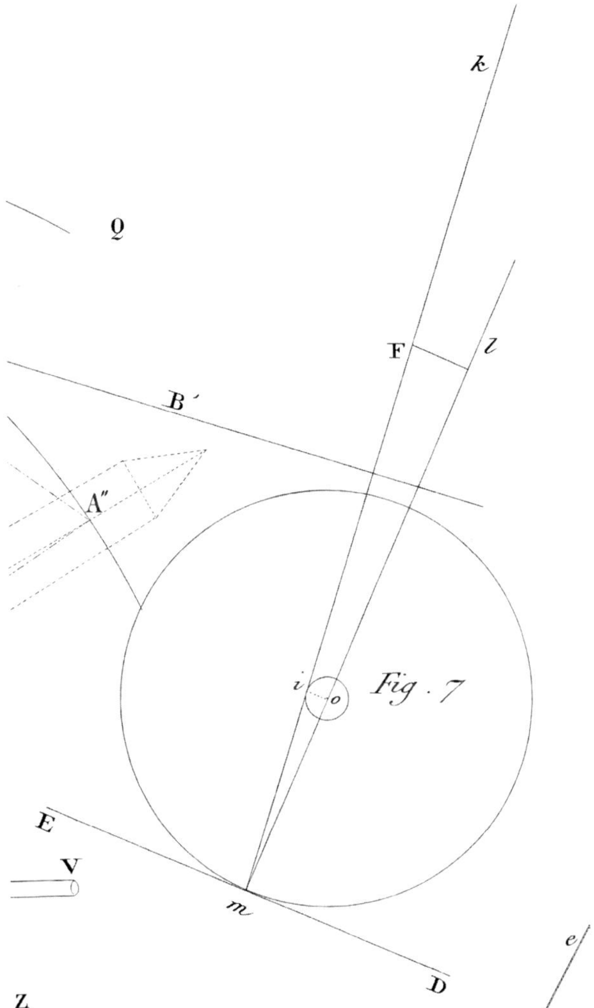

Detail from Plate XV, figure 7: Representation of the friction and tension created by the weight of the obelisk

laborious. In order to simplify them, we will observe that one can remove from the preceding equations the terms $r\,f'$ R and $r'\,f'$ R′. The first expresses the *moment* of the friction of the pivot of the obelisk; the second, that of the friction of the base of the cladding. The distances r and r', compared to the arms of the lever of the forces of the system, being very small, it follows that the quantities $r\,f'$ R and $r'\,f'$ R′ are negligible. It could even be demonstrated a priori that the tension to be transmitted to the resultant of the apparatus to overcome these resistances does not amount even to 0.1 kg of force exerted by each man on the ends of the bars. Other calculations are easy to determine with the following considerations: The system being in equilibrium, in whatever position, let (4) be the center of gravity of the mass. At the origin of the movement, this point is located in (1); the axis of the cladding is directed along the line o' B; the resultant of the cables coincides with the straight line CB whose length is invariable; the resultant of the apparatus coincides with BA, which abuts at fixed point A, where the fixed pulleys of the derricks are attached. This point is situated at the middle of the distance between the two extreme grommets.

During the rotation, the two straight lines oC and o(I) rotate around point o as the bars of a winch would do, where om would be the trunk; they always form among themselves an angle equal to Co(I), while their extremities (I) and G describe circumferences which have as a common center the point o. When the first coincides with *o* (4), the position of the second is determined by the condition that the arc $u'(4) = u(\mathrm{I})$. If one describes an arc of a circle from point (4′) as center with a radius equal to BC, this arc will intersect the circumference described by the point B around *o*′ in B′. Joining B′*o*′, B′(4′), and B′A will give the corresponding position of the axis of the cladding, of the resultant of the stays, and that of the apparatuses. That being established, the obelisk is held in (4) by its weight, which acts along the vertical (4)*l*, and in (4′) by the tension of the stays directed along (4′)B′. The resultant of these two forces thus passes through the intersection K of these two straight lines, and through the point of contact *m* of the obelisk with its support. Through this last point, let us draw a tangent DE to its circumference, which we have represented separately at a larger scale (figure 7 on Plate XV). It is obvious that the obelisk is on this tangent as the ship is on its launch gangway (see page 203). Thus if we break down the force F*m* into two others, F*l* and *ml*, one parallel and the other perpendicular to the tangent DE, the first one will be equal to the friction and tension determined by the pressure exerted by the second one on the inclined plane DE. We have seen that these resistances are equal to 1/12 of the pressure, thus F*l* = *ml*/12. Through point *o* let us draw a perpendicular *o*I on the straight line F*m* in the two right triangles *mo*I and *m*F*l*; we will have:

$$\overline{om}^2 \text{ or } r^2 \quad \overline{o\mathrm{l}}^2 + \overline{\mathrm{l}m}^2$$
$$o1 : 1m :: \mathrm{F}l : lm :: 1:12$$

in which by substituting and observing that $r = 0.20$ m, we have:

$$(0.2)^2 = \overline{o1}^2(1+12^2) \text{ and } \overline{o1} = 0.016\ \mathrm{m}$$

Thus, the resultant *km* (figure 2) is tangent to a circumference described from point *o* as a center with a radius equal to 0.016 m. The surface of the circle measured on the scale of the drawing can only be represented by a point of a compass. In other words, the resultant *km* is materially the same as the line *ko*. Since the tangent of the angle that these two straight lines form is proportional to the moment of the friction, this resistance is negligible. This said, let us agree to represent by the unit of length a force that pulls 20,000 kg. And let us take, starting from point *k*, a distance *ke* = 250,000 kg, or the weight of the obelisk. The resultant of the two forces *ke* and kB′ being directed along *ko*, let us construct the parallelogram of the forces by bringing through point *e* a parallel to (4′)B′; *e*G will represent the tension of the stays and *k*G the size of the resultant, or the pressure sustained by the obelisk. Let us bring this length from *o* to *k*′; the distance from this point to the vertical axis *oy*′ will be the horizontal component of this force, or the horizontal effort which tends to displace the upper surface of the pedestal, and *k*′*k*″ the vertical pressure exerted on the obelisk.

The cladding is completely similar to the obelisk in terms of equilibrium. It is supported at its top B′ by the force B′*a* = *e*G = the tension of the stays, and on the other side by that of the apparatuses. We wish to demonstrate, as previously, that the resultant of these two forces is materially equal to B′*o*′. Drawing the parallelogram of the forces, *ab* parallel to B′A will represent the tension of the apparatuses, and B′*b* the pressure exerted on the base of the cladding. Bringing this length from *o*′ to *b*′, the distance from this point to the two rectangular axes, *o*′*y* will measure the horizontal pressure and *o*′*x* the vertical pressures, supported by the hinge. Similarly, if we bring in the direction AB, and starting from point A, a length Ar equal to *ab*, and if through this point we draw the perpendiculars *r*A″ and AA″, the first will give the measure of the force of vertical traction, and the second that of the force of horizontal traction exerted on the fixed point of the derricks. Finally, the distance *or*′ = *ba* drawn on the line joining the center of action of the point of attachment of the stays will represent, in size only, the resultant of the tension of those cables.

Examination of the figure will show that there are 20 of them; they intersect 2 by 2 on the straight line B′(4′), following the angles that this line divides into two equal parts. Let *x*A*y* (figure 8) represent the average of those angles, A*p* the line that divides the angle into two equal parts. Let us take, starting from point A, a distance *Ap* equal to *or* and let us draw the parallelogram of the forces; A*x*/10 will be the tension common to each stay.

Working in the same manner, when the center of gravity of the mass is found in (1), (2), (3), (4), (5), (6), (7), four groups of points or a geometric plane are obtained, with which the magnitude of all the forces for any position of equilibrium can be determined.

It is evident that everything we have just said applies equally to the lowering apparatus. Thus, we will confine ourselves to indicating the constructions that will allow us to know the measure of the tension of the lowering cable.

The cladding is supported at its center of gravity *o*″ by its weight, 22,400 kg, acting along the vertical *o*″*e* . It is maintained in its equilibrium on its pivot by the stays that pull along RA′. The resultant of these two forces thus coincides with A′*o*′. If we construct the parallelogram of the forces, drawing from point A′ a distance A′*c* = 22,400 kg, *cd*′ parallel to A′ R will be the tension of the stays. Let us take A′G′ = 250,000 kg, or the weight of the mass of the granite, and combine into one single force the three forces acting in A′, and whose resultant is directed along A′*o* since the obelisk is in equilibrium. To this end, let us draw first the parallelogram A′*c*′*a*′G′ in which A′*c*′ is equal to *cd*′, then through the point *a*′ let us draw a line parallel to A′B′; the distance *a*′*a*″ will measure the tension of the lowering cable.

The results deduced from these graphic constructions, and done with precision, will differ only in negligible quantities from those that we would have obtained by using the preceding formulae. This is how we constructed the following tables.

Retaining Apparatus

(Measurements in kilograms)

Position of the center of gravity of the obelisk	1	2	3	4
Resultant of stays	33,709	61,238	82,840	100,916
Tension on each	2,142	3,192	5,270	6,416
Resultant of apparatuses, or traction exerted on the fixed points of the derricks	32,000	56,620	76,000	95,040
Vertical traction on this same point	30,300	51,800	64,500	72,000
Each mobile pulley is pulled from bottom to top as follows:	4,000	7,080	9,500	11,880
Tension transmitted to the sleeper, which, in this case, supports the strongest load expressed by $T_1 = 0.245 \cdot M$	980	1,735	2,327	2,910
Tension transmitted to the extremity of the chief pulley, calculated from the formula $u = 0.001084$ M, or the force to be exerted by each man to balance the system	4.34	7.67	10.30	12.88
Pressure on the body of the obelisk	263,000	260,800	241,600	211,000
Horizontal component of this pressure	31,000	61,500	79,600	82,000
Pressure on the body of the cladding	29,600	62,000	94,000	129,000
Horizontal pressure that tends to move it	21,800	37,600	38,200	20,000

Erecting Apparatus
(Measurements in kilograms)

Position of the center of gravity of the obelisk	1	2	3	4	5	6	7
Resultant of stays	124,000	121,800	120,000	114,000	100,000	76,000	42,400
Tension on each	6,302	6,190	6,098	5,793	5,082	3,862	2,155
Resultant of apparatuses, or traction exerted on fixed Point A	104,000	89,600	88,000	86,200	80,000	64,700	38,000
Vertical traction on the same point	56,300	55,700	62,900	66,300	62,500	53,700	31,500
Pressure on the body of the obelisk	126,000	140,000	183,800	231,200	262,700	276,600	274,400
Horizontal component of this pressure, or effort, or force that tends to displace the plinth of the pedestal	0,000	36,800	83,600	105,900	99,900	72,400	35,900
Pressure on the body of the cladding	200,900	152,600	152,600	123,000	90,000	51,000	19,200
Horizontal pressure that tends to move it	85,900	31,780	23,800	50,600	53,300	38,200	16,000

To the explanations that we have already given about each of these tables (see pages 60, 61, 62, 163, and 164), we have to add the calculation relating to the force to be generated by the men on the bars of the 10 capstans in order to overcome the resistance transmitted by the weight of the monolith to the stays, and through them to the ropes of the derricks.

The capstans are 0.80 m in diameter; the bars on which the men push are 16 in number, each of them pushed by three men, which makes 48 per capstan and 480 total. Let us calculate from this number. The average radius of the bar each man pushes is 2.45. By estimating the force of a man at 12 kg, the effect transmitted to the rope that wraps around the capstan will be given by the formula 2RF = 0.927 T in which F = 12 kg × 48 = 576 kg and R = 2.45 m; thus T_n = 3.044 kg. The return pulley, which changes the direction of the rope at a right angle, absorbs 15% of it [formula (6) (0.85) T = X], thus leaving 2,587 kg.* The pulley block having 7 branches, we have (see the force of the motors):

$$T_7 = 0.225 \cdot M = 2.587 \text{ kg thus } M = 11{,}500 \text{ kg}$$

The 10 capstans will thus pull the upper tie beam of the cladding as if it were 115,000 kg, while the force transmitted by the mass to the stays and through them to the cords of the derricks is 104,000 kg at the beginning of the motion and diminishes gradually (apparatus of erection, resultant of apparatuses). Since a vigorous man can produce a force of 20 kg instead of the 12 I had assumed, we can surmise that 300 men could have been sufficient for the operation.

Nothing more remains to complete this work than to examine whether the ropes of the system are adequate to resist the forces that they will have to support, and to compute their absolute strength. Duhamel did a great deal of work on this, but the results he obtained cannot be applied to ropes used today. The process of rope making has been improved by Mr. Hubert, director of naval construction. The new mode of manufacture means that the strength of a rope is proportional to the number of strands forming it, which is the same as the square of its diameter or its circumference. This translates algebraically to the formula:

$$F = mc^2 \begin{cases} F \text{ is the strength of the rope in kilograms} \\ c \text{ is the circumference in centimeters} \\ m \text{ is a constant to be determined by experience} \end{cases}$$

Given this, the test performed on the hemp ropes before supplying them to the arsenals consisted in building a 40-ply rope of 47 mm, which had to support a weight of 1,600 kg without breaking. Inserting these two numbers in the preceding formula, one finds that m = 72.7. But the 40-ply rope, which supports this weight for only a few seconds, would snap under the continuous action of a lesser weight. It would therefore be imprudent to come too close to this limit. Nevertheless, one can agree that it would resist for several hours a tension half of that which would cause it to break. For greater safety, and in order to allow for unforeseen circumstances, we reduced the weight that could be put on a rope for a long time without danger of breaking to about 1/3 of the weight under which the 40-strand rope will break, according to the "roman scale" measurement of hemp strength. This weight is calculated from the formula:

$$F = 25\, c^2$$

The circumference of a stay is equal to	28.27 cm	thus $c^2 = 799.19$
That of the tackles is equal to	15.7	thus $c^2 = 246.49$
Absolute weight on the first rope	19.980 kg	
Absolute weight on the second	6,162	

* In Fontana's apparatus the chief of each derrick went through five return pulleys before winding around the capstan. In this arrangement, more than half of the force transmitted to this rope by the actions of the men and horses was absorbed by the friction and stiffness of the rope.

It follows from the preceding calculations that the absolute strength of these ropes was double that of the weight that they had to support in all the maneuvers done on the monolith. Thus, all that would have been needed was a stay of 20 cm and a tackle of 12 cm in circumference. These cables would have supported the forces indicated in the two tables, without breaking. But let us observe again here that the results deduced from the theory can be considered only as lower limits. Indeed, one assumes that the ropes of the system are equally tense, that the pulley blocks unwind in a uniform manner, that there is neither deformation nor compression, that the surfaces in contact are rabbeted and polished through long use, etc., etc., etc. But not all these assumptions are met in practical applications. This results in greater or lesser losses of strength, shocks that could break the apparatuses were they not made of exceptional materials, executed with precision, and large enough to resist a force almost double what had been calculated without taking all these difficulties into account.

A rare portrait of Apollinaire Lebas, who died on New Year's Day, 1873 (from *Le Monde illustré*, January 11, 1873)